HAZWOPER
Incident
Command

A Manual for
Emergency Responders

David M. Einolf

GOVERNMENT INSTITUTES
An imprint of
THE SCARECROW PRESS, INC.
Lanham • Toronto • Plymouth, UK

Published in the United States of America
by Government Institutes, an imprint of The Scarecrow Press, Inc.
A wholly owned subsidiary of
The Rowman & Littlefield Publishing Group, Inc.
4501 Forbes Boulevard, Suite 200
Lanham, Maryland 20706
www.govinstpress.com

Estover Road, Plymouth PL6 7PY, United Kingdom

British Library Cataloguing in Publication Information Available

Library of Congress Cataloging-in-Publication Data

Einolf, David M.
HAZWOPER incident command: a manual for emergency responders / by David M. Einolf.
        p.   cm.
    Includes index.
    ISBN: 0-86587-612-6
    1. Hazardous waste site remediation—United States—Safety measures. I. Title.
TD1050.S24E36   1998
363.72'87'0973—dc21
                                   97-52719
                                    CIP

Table of Contents

Preface

I believe that no OSHA standard is more misunderstood than the HAZWOPER standard as it is applied to industrial facilities. Early, unscrupulous course providers sold unnecessary training to unsuspecting clients who later reacted by eliminating all training in the area of emergency response. The text which follows is an attempt at moderation. It is not a text for HAZMAT Technicians or First Responders. I firmly believe that quality texts in this area exist and should be used.

This book is a text for training Facility Incident Commanders. By this I mean personnel who are responsible for selecting, training, and outfitting facility responders who will fill any of the first three HAZWOPER training roles: First Responder Awareness, First Responder Operations, and HAZMAT Technician. The Facility Incident Commander (IC) bears a tremendous burden. Strictly interpreted, he is responsible for the health and safety of all employees serving on the HAZMAT team. This text serves to help the IC complete this important task. Through a combination of examples, forms, and useful materials, the book weaves the fabric necessary to support the IC in an industrial setting.

I hope that this text will be used by trainers of facility ICs. I have assembled materials I feel will be useful in such a course, obviating the need to gather materials from diverse sources. The manual is meant to provide a quick reference guide for facility ICs without inundating them with material. The goal of the text is to provide a safe working environment for facility HAZMAT personnel through support of their commander. The text integrates both Process Safety Management (PSM) and Accidental Release Risk Management Planning (RMP), important components in today's changing regulatory climate.

David M. Einolf
Portland, Oregon
January 1998

About the Author

David M. Einolf is a Principal of Dames & Moore, Inc. and Manager of the Pacific Northwest Compliance and Operations practice. Mr. Einolf also serves as the firm-wide coordinator for process safety management (PSM) and accidental release risk management (RMP) services for Dames & Moore. Mr. Einolf has developed PSM and emergency response programs for numerous United States companies, including a number of *Fortune 500* clients. Among notable assignments, Mr. Einolf assisted Kraft Foods in the development of a firm-wide emergency response plan; GTE, in the development of a hazardous materials transportation training program for all line employees; Tenneco Packaging, in the development of spill prevention, control, and countermeasures (SPCC) plans for all United States facilities; and Albertson's, Inc., in the development of comprehensive Process Safety Management (PSM) programs.

Mr. Einolf has worked with numerous food processors throughout the United States in the development of PSM programs and has conducted numerous HAZWOPER training sessions. He has trained both public and private sector employees in all aspects of hazardous materials handling. Mr. Einolf holds a B.A. in Earth and Planetary Sciences from the Johns Hopkins University, a M.S. in Marine Biochemistry from the University of Delaware, and a M.B.A. from the University of Chicago. Mr. Einolf resides in Portland, Oregon, with his wife and two children.

Acknowledgments

I would like to acknowledge the contributions of all of my clients and students over the years who have provided me with the opportunity to hone my craft and to learn new materials. I would especially like to thank Kim Braegger, Cynthia Forsch, and Walt Bentley, each of whom listened to the same bad jokes in many separate sessions. I am especially grateful to Dames & Moore for allowing me to complete this manuscript and pursue my profession with a remarkable level of support.

Portions of this text were developed for commercial clients as well as the American Industrial Hygiene Association (AIHA), the Pittsburgh Conference (PittCon), and HazMat World (now EMAT) whose support is appreciated.

On a more personal basis, I'd like to thank George Krafcisin for believing, and all of my support staff throughout the years. Special thanks are reserved for Becky, Allison, and Davis Einolf for keeping the home fires burning when I was out training in the Ozarks, Chile, or Africa.

1

DEVELOPMENT OF THE HAZWOPER REGULATION

INTRODUCTION

OSHA Standard 29 CFR 1910.120: Hazardous Waste and Emergency Response Operations, or HAZWOPER, was placed into effect in late 1989 amid growing concern over the activities and dangers of unregulated cleanup activities at hazardous waste sites.

Since the passage of the Comprehensive Environmental Response, Liability and Compensation Act (CERCLA) in 1980, hazardous waste cleanups had accelerated throughout the United States. A combination of federal (Superfund), state, and private efforts increased activities at some of the most egregious hazardous waste dump sites. As a result of this increased activity, more workers were needed to effect these cleanups.

The workforce for the traditional CERCLA cleanup was drawn from those groups who had access to the equipment needed to complete the task, namely, the construction trades. On the whole, the employees of these construction contractors had little or no familiarity with hazardous materials handling, personal protective equipment, or safe work procedures.

This untrained workforce led to three problems:

- Worker exposure to hazardous chemicals
- Workplace accidents related to hazardous chemical handling, including serious explosions and fires
- Additional releases of hazardous materials from work sites with potentially far-reaching consequences

In response to a growing number of such incidents, and a growing number of worker injuries from such incidents, both OSHA and the EPA were asked by Congress to consider regulations to cover the activities of workers at hazardous waste cleanup sites.

During the public comment on OSHA's proposed standard, OSHA was petitioned to expand the scope of its regulation to operating facilities in which employees might be asked to cleanup (remediate) spills of hazardous materials. In response to these comments, OSHA added paragraph (q) to the regulation, which focuses largely on worker protection at operating or manufacturing facilities.

THE HAZWOPER STANDARD AND OPERATING FACILITIES

HAZWOPER has two basic areas of impact on operating facilities: planning and training. In the planning category, HAZWOPER requires that facilities have an emergency response plan which contains the following components:

- Pre-emergency planning and coordination with outside agencies
- Personnel roles, lines of authority, training, and communication
- Emergency recognition and prevention
- Safe distances and places of refuge
- Site security and control
- Evacuation routes and procedures
- Decontamination
- Emergency medical treatment and first aid
- Emergency alerting and response procedures
- Critique of response and follow-up
- Personal protective equipment and emergency equipment
- Usage of local and state plans

In terms of training, HAZWOPER mandates that training be provided at five levels, depending on the type of response which the facility wishes to undertake. Those five levels are prescribed as follows:

- First responder - awareness level
- First responder - operations level
- Hazardous materials technician level
- Hazardous materials specialist level
- Incident commander level

Facilities can provide three levels of response under the HAZWOPER standard:

- Evacuation
- Containment and use of an outside HAZMAT team

- In-house response using a HAZMAT team

For each of these three levels, OSHA prescribes levels of training. Simply put, the three levels correspond to the first three training levels:

- Evacuation: First Responder Awareness
- Containment: First Responder Operations
- Control: HAZMAT Technician

Exact numbers and needs for an individual facility are matters left for a careful reading of the standard, OSHA's compliance guideline, CPL 2-2.59, and as a decision for the facility emergency response coordinators.

The focus of this book is to provide a framework for facility coordinators (on-scene incident commanders) to make decisions regarding emergency response. The book reviews the Incident Command System, emergency response planning, and specialty skills required of the incident commander.

HAZWOPER AND SITE OPERATIONS

The bulk of the HAZWOPER standard is directed towards employees of firms involved in the commercial (or municipal) cleanup of hazardous materials. Such cleanups include U.S. EPA-directed cleanups of national priority list (NPL) or Superfund sites; state "voluntary" cleanups under the Resource Conservation and Recovery Act (RCRA); and activities at hazardous waste treatment, storage, and disposal (TSD) sites.

The standard spells out requirements for site safety and health plans to be followed at such sites. Written programs for such sites have the following components:

- An organizational structure
- A comprehensive workplan
- Site specific safety and health needs
- A safety and health training program
- A medical surveillance program
- Standard safety and health operating procedures

Components from these plans are an important part of the requirements of the facility emergency response plan. For instance, facilities with trained emergency response teams are required to have medical surveillance programs substantially similar to those of professional emergency responders.

The HAZWOPER standard also provides an interface between many other OSHA regulatory programs. A site supervisor and a facility incident commander must have an understanding of a number of OSHA programs:

- Personal protective equipment (29 CFR 1910 Subpart I (1910.132-138)
- Hazard communication (29 CFR 1910.1000)
- Confined space entry (29 CFR 1910.146)

- Lockout/tagout

APPENDICES TO THE HAZWOPER STANDARD

The HAZWOPER standard, as presented in the Code of Federal Regulations, contains five "non-binding" (meaning that they cannot be enforced) appendices:

(A) Personal protective equipment test methods

(B) General description and discussion of the levels of protection and protective gear

(C) Compliance guidelines

(D) References

(E) Training curriculum guidelines

These appendices contain practical documentation for the facility incident commander. Appendix A provides a guideline for testing totally-encapsulating chemical protective (TECP) suits. Appendix B provides the guidance for levels A-D of protective clothing. Appendices C and E provide compliance information and a detailed description of suitable programs for all levels of HAZWOPER training.

OSHA has considered developing a standard (tentatively, 29 CFR 1910.121) for hazardous waste site worker training programs, but has not done so, even after almost ten years of HAZWOPER regulation.

2

IMPLEMENTING THE FACILITY INCIDENT COMMAND SYSTEM

INTRODUCTION

The concept of an incident command system was developed by federal fire training organizations (the National Fire Academy and others). It is designed to function for small and large organizations and lead to the efficient application of resources to difficult and potentially dangerous situations.

While the system was developed for fire protection, the EPA and OSHA have integrated the concept of the ICS into hazardous materials protection. The EPA, through the mandate of SARA Title III, has developed incident command structures at the federal and state level. State Emergency Management Agencies have called upon local agencies to develop similar programs. The ICS is flexible and can be used by both small and large organizations. This flexibility is achieved by adding and subtracting modules as needed, responding to the size of the facility and the size of the emergency.

COMPONENTS OF THE INCIDENT COMMAND SYSTEM

The National Inter-Agency Incident Management System (NIIMS) enumerates the following components of an incident command system:

- Common terminology
- Consolidated action plans
- Modular organization
- Manageable span of control
- Integrated communications
- Predesignated incident facilities
- Unified command structure
- Comprehensive resource management

COMMON TERMINOLOGY

Incident commanders have developed a standard terminology for dealing with hazardous materials responses. This terminology is detailed in the National Fire Protection Association (NFPA) Standard of Practice 471: "Responding to Hazardous Materials Incidents."

Through the use of a common terminology, terms such as "control zone" or "hazard sector" will have the same meaning to responders in Albany, NY, and Albany, MN. In addition, coordination between agencies is preserved through the common vocabulary.

Modular Organization

The incident command system is broken into modules which can be expanded depending on the role being played during the incident, or during the planning stage. In addition, the modular organization ensures some consistency of personnel roles across organizations. Once again, the modular organization is useful in coordinating incidents between agencies.

Integrated Communications

The ability to communicate (or the lack of communication) is a major factor in hazardous materials incident response. An established ICS should provide methods for ensuring that the most effective means of communication is used. In addition, specific protocols for the use of various types of communications equipment should be established.

Most communities have well established emergency communications frequencies and programs. Emergency channels are opened for a particular response and are controlled by a central group.

Lastly, communications centers on the development of a standardized set of signals or a set of procedures for radio, telephone, or visual communication. Multi-agency operations have done away with the use of coded ("10" codes, etc.) transmissions to reduce the confusion between operating parties.

Unified Command Structure

Hazardous materials emergencies generally involve several agencies acting in coordination. In such efforts, all agencies must be given the opportunity to participate in all decision-making efforts. Despite this cooperation, decisions resulting in control of the operations must emanate from a central point, hence, a unified command.

Consolidated Action Plans

A successful emergency response is a well-planned one. While the combination of planning and emergencies may seem strange, agencies expected to respond to a hazardous materials emergency will work together to develop a comprehensive local plan. An effective facility incident commander will work with the local agencies to ensure that his facility is included in the program and that the responsibilities of the responders are outlined in sufficient detail to prevent duplication of effort.

Manageable Span of Control

A time-honored truth in management is that a person's span of control is limited. The number of direct subordinates that can be managed at any one time is generally limited to seven persons. In establishing an incident command system, care must be taken to limit the number of direct reports in any one position to five.

Using the modular ICS, the organization can expand and managerial responsibilities can be delegated down the organizational tree as the size of the response increases. This distribution of responsibility ensures that no person will be overwhelmed by the requirements of the position.

Predesignated Incident Facilities

The cornerstone of the incident command system is the establishment of a designated facility for dealing with the response. Typically, an organization will designate a "command post," which serves as the communications center point for the response, and a series of "staging areas" which are assembly points for the emergency responders. Command posts are usually established in areas which are likely to be out of the danger zone of an anticipated incident (e.g., upwind, away from chemical storage and handling). Staging areas are often stocked with emergency response equipment and may even be suitable for the establishment of a mobile command post.

Comprehensive Resource Management

The ICS provides centralized command and control functions for an incident. Use of the ICS allows multi-agency/multi-jurisdictional responses to be carried out in a safe and effective manner.

FUNCTIONAL AREAS OF THE ICS

It is possible to divide the ICS into five areas of responsibility: command, operations, logistics, finance, and planning. Incident command structures established for small communities and facilities may have some or all of these functions integrated into a single position. On the other hand, emergency response activities for a community of several million (such as Chicago or New York City) would involve coordination of several people at each level.

The *command* position is responsible for coordinating the response operation. The incident commander is the person responsible for the command function. The incident commander must assess the situation, establish goals and procedures to control the situation, and allocate resources to complete the emergency response. Each of the other functions is subordinate to the command function.

Operations is responsible for all tactical operations used to bring the emergency under control. All procedures to control, confine, and contain releases are handled by the operations function.

Equipment, supplies, and services needed to complete the emergency response are provided by the *logistics* function. Logistics includes stocking the appropriate materials, ensuring rapid and adequate supply, and restocking in the aftermath of the incident.

Planning and *finance* provide information and financial resources needed to create and maintain and effective emergency response organization.

COMMAND STRUCTURES AND THE DELEGATION OF COMMAND

Two forms of incident command structure exist: single and unified. Single commands exist when only one organization is involved in a response. As an example, suppose a facility with a fully trained hazardous materials response (HAZMAT) team suffers a release of anhydrous ammonia from a refrigeration system. If this release is confined to the premises and the local emergency response organization is not involved in the response, then the incident has been controlled by a single command.

On the other hand, if the facility suffers an uncontrolled release which threatens to move off the property and endanger the local population, the team may need to involve other local emergency response personnel. At this point, the incident command system will be under unified control. That is, all emergency responders at the scene will be under the command of a single incident commander.

In most cases, the local fire department will take command of any incident in which it is involved. Well developed procedures exist in most jurisdictions which give responsibility to the fire department. Larger responses may be headed up by officials in the state or federal emergency management organization.

Whatever the command structure, the importance of establishing a chain of command in advance of an incident is paramount. There is no time for controversy or confusion surrounding who should control the resources available to mitigate a hazardous materials response. Such conflicts should be alleviated through pre-emergency planning with the local emergency response commission. Each facility should have a detailed delegation plan reflecting the delegation chain for varying types of emergencies.

PERSONNEL ROLES IN AN INCIDENT COMMAND SYSTEM

Safety Officer

Under the HAZWOPER standard, facilities with a duly constituted hazardous materials team must have an established site emergency safety plan. Such a plan must establish the bounds of safe conduct to be met when undertaking an emergency response. As such, each facility should have a designated safety officer as part of the HAZMAT team. The safety officer will be responsible for receiving a briefing from the incident commander in order to execute the following tasks:

- Identify hazardous situations related to the incident
- Review incident action plans
- Identify potentially unsafe situations
- Exercise emergency authority to stop and prevent unsafe acts
- Investigate accidents that occurred within the incident area
- Review and approve the medical plan
- Maintain a HAZMAT team safety log

Public Information Officer

In any crisis situation, there is a demand for fast and accurate information concerning the crisis. This holds equally true for hazardous materials incidents. The Public Information Officer (PIO) has the responsibility to control the location of the press and to provide accurate statements concerning the progress of the incident.

The PIO should establish a press area, well away from any potential hazards and in no danger of jeopardizing the ongoing operations. A detailed procedure for press interaction should be obtained by the PIO or the incident commander during the development of the facility's emergency response plan. Most companies have a program for dealing with public relations in disaster situations.

The importance of an employee's role in dealing with the press should never be underestimated. The public regards hazardous material incidents as only one step less dangerous than nuclear material releases. Improper press responses or a lack of attention to the hazards can result in a far greater disaster than the actual release.

Liaison Officer

Many larger facilities will consider having a liaison officer as part of the incident command staff. As with the other incident command positions, the liaison officer may serve as a backup incident commander. The liaison officer serves to coordinate the facility's interaction with public sector organizations such as fire and police departments, hospitals, and the local Red Cross.

BASIC STEPS TO INCIDENT COMMAND

Command of an incident has three basic steps:

- Assume command
- Establish control
- Establish a command post

Assume Command

Generally the incident commander will be assuming command from a first responder at the awareness or operations level. The first responder will have performed some of the duties under the facility response plan. It is likely that the first responder will have evacuated the immediate area and may have contained the spill to a smaller area.

The incident commander will need to debrief the first responder and obtain information concerning the size, location, and nature of the spill; nature of the operations endangered by the spill; and possibility of incompatible materials near the location of the spill.

The incident commander will then need to take control of the incident response from the first responder.

Establish Site Control

The site control process must be put into place quickly if there is a threat of contamination or if contamination exists. Lines of demarcation between the hot and warm zone must be established, access to the site must be limited, and a cold zone must be marked off.

Establish a Command Post

The incident commander must retain control of operations until he is relieved by someone at a superior level of response. A command post should be created which will allow for this smooth transition of command. The command post should be located at the edge of the warm zone at the end of the decontamination line (or contaminant reduction corridor).

3

IMPLEMENTING THE FACILITY EMERGENCY RESPONSE PLAN

INTRODUCTION

Aside from the training considerations mentioned in Chapter 1, the major impact of OSHA's HAZWOPER standard is the additional burden of emergency planning for facilities.

Under the regulation, each facility is required to prepare a detailed plan which addresses, at a minimum, the following areas:

- Pre-emergency planning and coordination with outside agencies
- Personnel roles, lines of authority, training, and communication
- Emergency recognition and prevention
- Safe distances and places of refuge
- Site security and control
- Evacuation routes and procedures
- Decontamination
- Emergency medical treatment and first aid
- Emergency alerting and response procedures
- Critique of response and follow-up
- Personal protective equipment and emergency equipment

While the standard does not require the integration of the HAZWOPER plan with other federal- and state-mandated emergency response plans, many facilities have opted for inclusion of plans prepared under

SARA Title III, RCRA, the Oil Pollution Act of 1990, and the Clean Water Act in their HAZWOPER plans.

The discussion which follows reviews each of the requirements of the HAZWOPER plan, taking into account the actions needed for an incident commander to implement the plan.

PRE-EMERGENCY PLANNING

The incident commander is responsible for coordinating emergency response planning with all applicable local, state, and federal agencies. In addition, a comprehensive emergency response plan should include detailed contact lists for a range of outside services, from electrical and HVAC contractors to providers of backup electrical generators.

The incident commander should contact all of the public agencies that would be affected in an emergency:

- Fire
- Police
- Civil Defense (Emergency Management)
- Hospitals

Each of these agencies should be made aware of the activities ongoing at the facility and of the hazardous chemical usage at the facility. While the facility has (if it is in compliance with SARA Title III) submitted information concerning chemical usage and storage to local fire departments, it is unlikely that it has followed up with the fire department coordinator to determine if there are any questions concerning use or emergency response.

A listing of emergency contacts is included as Appendix A.

The incident commander should send a copy of the facility emergency response plan to the appropriate contacts at each public agency.

PERSONNEL ROLES, LINES OF AUTHORITY, TRAINING, AND COMMUNICATION

It is the responsibility of the incident commander to ensure that each participant in an emergency response possesses a clear understanding of his or her role, as well as those of the immediate supervisor and subordinate, and annual training requirements.

In order to further clarify the role of a particular emergency responder, many planners find it convenient to develop an emergency response plan which is individualized for each employee. Each person receives a sheet which lists his or her responsibilities under all circumstances. All of the sheets are then integrated into the total emergency response plan.

Each emergency response plan should have a detailed organization chart which outlines the responsibilities of every person within the facility. Contact lists, telephone trees, and response numbers should be provided to each responsible party.

Records of the type, number, and content of training sessions should be kept with the emergency response plan for ease of inspection. Training records should be kept for every employee who has a designated duty within the plan. A sample training record is included as Appendix B.

The lines of communication for the facility should be outlined in the plan. Appropriate contacts (on a 24-hour basis) should be listed for the facility. Contacts should be at the appropriate levels to make reports to local, state and federal authorities. Failure to make a timely report of a spill or leak could lead to a significant fine or criminal penalty. A spill notification form is included as Appendix C.

EMERGENCY RECOGNITION AND PREVENTION

As with a spill prevention, control, and countermeasures (SPCC) plan, a detailed inventory of chemical hazards within a facility should be performed. Such an inventory can be performed as part of SPCC planning, process safety management, or as part of HAZWOPER preparation. The aim of the inventory should be to indicate the likely locations of chemical hazards and the steps taken to mitigate potential accidents at those locations.[1]

Facilities which use specific warning signs, alarms, sirens, or codes should detail those practices in the emergency response plan.

SAFE DISTANCES AND PLACES OF REFUGE

Plans developed to conform with HAZWOPER should include evacuation plans substantially similar to those required by OSHA under 29 CFR 1910.38(a). Special care, however, should be taken to calculate safe distances based on worst-case releases of hazardous materials, rather than limiting the scenario to fire or similar emergencies.

Airborne materials, such as gaseous chlorine or anhydrous ammonia, represent special hazards in emergency situations. They require an understanding of the dispersion of the chemicals in order to determine safe distances and location of places of refuge. Facilities engineers, vendors, or trade associations should be consulted in developing such evacuation plans.[2]

The EPA has recently (June 20, 1996) issued Final Rules (40 CFR 68) requiring facilities using hazardous materials above a certain thresholds to develop facility accidental release risk management plans (RMPs) (see Appendix E). The regulation also requires managers of such facilities to analyze the effect on surrounding areas (off-site consequence analysis) of a hazardous-materials release from facility operations.

---

[1] Guidelines for the preparation of a SPCC plan are not widely available. Plans prepared in accordance with the U.S. EPA guidelines may not be suitable to other (local) jurisdictions. The guidelines developed by the Metropolitan Water Reclamation District of Greater Chicago (MWRDGC) are included as Appendix D.

[2] A number of computer programs exist to model the effects of spills of hazardous materials. Assistance with these models (CAMEO, SLAB, ARCHIE, DEGADIS) may be available from the local emergency planning committee or through the state.

Compliance with 40 CFR 68 will require facilities to determine the safe distances from a worst-case release of hazardous materials. The rule requires that this information be published in an electronic form and be accessible to the public.

SITE SECURITY AND CONTROL

It is the responsibility of the incident commander to develop plans for restricting access to the site of a hazardous-materials incident. These plans can include specific procedures for the establishment of hot/warm/cold zones, contamination reduction corridors (CRC), and command posts.

Since different chemical releases have different effects, care should be taken to accurately determine safe distances for site control for each possible disaster scenario.

This section of the plan might also include detailed information on the numbers and types of resources for security and control and should outline the chain of responsibility for establishment of control.

EVACUATION ROUTES AND PROCEDURES

The facility evacuation routes and procedures can also be those developed in compliance with 29 CFR 1910.38(a). Care must be taken to analyze the workplace to determine if any of the traditional fire evacuation routes might be endangered during a hazardous materials release. If so, those routes must be altered accordingly. This section of the plan should include a detailed facility route map with information showing which working groups will be using each path of evacuation. Personnel responsible for leading evacuation teams should be named in the Emergency Response Plan.

The RMP rules (40 CFR 68) require that a facility fully analyze the evaluation routes to be used within the facility and by the surrounding community. Through the development of the off-site consequence analysis, the community will learn the potential consequences of a spill or release at the facility.

DECONTAMINATION

Decontamination is critical to the health and safety of responders during hazardous materials incidents. It serves to protect responders from hazardous substances that may contaminate and eventually permeate the personal protective equipment used during the incident.

It is the responsibility of the incident commander to develop a plan for the proper decontamination of employees responding to a hazardous materials incident. The plan should consider the range of factors which may affect the decontamination procedures:

- Nature of the contaminants
- Amount and location of contaminants
- Personnel exposure
- Potential of substances to penetrate or permeate the equipment
- Number of personnel in the decontamination zone
- Process (or movement) of personnel through the decontamination zone

- Methods of decontamination available
- Protection of responders from inadvertent contamination during decontamination

EMERGENCY MEDICAL TREATMENT AND FIRST AID

The impact of a disaster on the available resources of a medical facility is hard to imagine. Hazardous materials incidents have overloaded hospital trauma units and exhausted a community's available ambulance corps. They can result in significant loss of life. Proper planning can help to mitigate such a disaster. An understanding of the limitations of care available at local facilities and the nature of transportation and care at remote facilities can help build an effective response.

Cooperation between the incident commander and the local medical community improves the chances that a community will survive a hazardous materials incident with minimal impact. Facilities using specific hazardous materials should be willing to provide antidotes or specific abatement materials to medical practitioners as donations or at a reduced cost.

Facility emergency personnel must be acquainted with the hazards and traumas that may be experienced in an incident.

EMERGENCY ALERTING AND RESPONSE PROCEDURES

Volunteer fire departments throughout the United States are fairly effective. Most have relatively simple systems for alerting members and directing them towards the incident. By the same token, a facility must establish a simple way to alert and muster the emergency response team. These procedures must be outlined in detail in the plan.

If specific warning sirens, bells, or plant whistles are used to indicate different types of emergencies, those sounds should be detailed in the plan and practiced on a regular basis so that the employees in the plant are familiar with them.

Some larger manufacturing plants have telephone numbers which have the various emergency alerting sounds recorded. Employees may use these numbers to quickly sound an alert.

CRITIQUE OF RESPONSE AND FOLLOW-UP

Regardless of the outcome of the response, a detailed critique should be undertaken. An evaluation form should be set up and detailed information obtained from every member of the team who was involved in the response.[3]

The incident commander is responsible for reporting the incident to the appropriate corporate, local, state, and federal authorities.

---

[3] A sample critique form is provided as Appendix G.

Lastly, the follow-up to the incident should include a review of the inventory of emergency response equipment and a restocking of missing equipment. Members of the response team should be polled to determine if their equipment was adequate for the response, and all inadequacies should be addressed.

PERSONAL PROTECTIVE EQUIPMENT AND EMERGENCY EQUIPMENT

The emergency response plan should include a detailed inventory of the equipment available to on-site responders including standard operating procedures for review, repair, and replacement of equipment.

Equipment-use, maintenance, and training logs should be maintained, as well as documentation of fitting sessions for respiratory protective equipment.

4

HAZARDS ASSOCIATED WITH EMPLOYEES IN CHEMICAL PROTECTIVE CLOTHING

INTRODUCTION

Personal protective equipment (PPE) is the designation given to the complement of clothing issued to an emergency responder pursuant to his level of training. PPE ranges from safety glasses, gloves, and steel-toed shoes to a fully encapsulating chemical protective suit.

The incident commander directing the activities of a HAZMAT team is responsible for the proper selection of personal protective equipment for each member of that team. It is the incident commander's responsibility to ensure that the equipment is in good working order and that the employee has received accurate and up-to-date information on the use of such equipment.

The incident commander, or his designee, is responsible for determining how long and under what conditions an employee can use personal protective equipment, and what personal protective equipment will be needed for a given contamination zone.

Given these responsibilities, the incident commander must make himself familiar with a full range of chemical protective equipment. He must understand the use of PPE and be able to select PPE given a specific chemical release. He must have an understanding of how to dress and undress (don and doff) using chemical protective clothing, and most importantly, he should understand the limitations of human physiology in light of the stresses induced in the use of chemical protective clothing.

LEVELS OF PERSONAL PROTECTIVE EQUIPMENT

The levels of personal protective equipment referred to in this chapter are described in detail in the appendix to the HAZWOPER regulation [29 CFR 1910.120 (Appendix A)].

Level A

Level A PPE is indicated when the greatest level of skin, respiratory, and eye protection is required. Level A protection should be worn when any of the following conditions exist:

- The hazardous material has been identified and requires the highest level of protection for skin, eyes, and respiratory system based on either the measured (or potential for) high concentration of atmospheric vapors, gases, or particulates, or the high potential for splash, immersion, or exposure to unexpected vapors gases, or particulates of material that are harmful to skin or capable of being absorbed through the skin; or

- Substances with a high degree of hazard to the skin are known or suspected to be present, and skin contact is possible; or

- Operations must be conducted in confined, poorly ventilated areas, and the absence of conditions requiring Level A has not yet been determined.

Level A PPE consists of the following gear:

- Pressure-demand, full-facepiece, self-contained breathing apparatus (SCBA), or pressure-demand, air-line supplied respirator with escape SCBA.

- Totally encapsulating chemical protective suit (TECP suit). A TECP suit means a full-body garment that is constructed of protective clothing materials; covers the wearer's torso, head, arms, and legs; has boots and gloves that may be an integral part of the suit, or separate and tightly attached; and completely encloses the wearer by itself or in combination with the wearer's respiratory equipment, gloves, and boots. All components of a TECP suit, such as relief valves, seams, and closure assemblies, should provide equivalent chemical-resistance protection.

- Coveralls *

- Long underwear *

- Gloves, outer, chemical-resistant

- Gloves, inner, chemical-resistant

- Hard hat (under suit) *

- Disposable protective suit, gloves, and boots (depending on suit construction, may be worn over totally encapsulating suit) *

- Two-way radios (worn inside encapsulating suit)

---

* This item is optional, but may be needed if conditions dictate.

Level B

Level B PPE is indicated when the highest level of respiratory protection is necessary, but a lesser level of skin protection is needed. The following conditions warrant the use of Level B PPE:

- The type and atmospheric concentration of substances have been identified and require a high level of respiratory protection, but less skin protection;

- The atmosphere contains less than 19.5 percent oxygen;

- The presence of incompletely identified vapors or gases is indicated by a direct-reading, organic vapor detection instrument, but the vapors and gases are known not to contain high levels of chemicals harmful to skin or capable of being absorbed through the intact skin; or

- The presence of liquid or particulates is indicated, but they are known not to contain high levels of chemicals harmful to skin or capable of being absorbed through the intact skin.

Level B PPE consists of the following gear:

- Pressure-demand, full-facepiece, self-contained breathing apparatus (SCBA), or pressure-demand, supplied-air respirator with escape SCBA

- Hooded, chemical-resistant clothing (overalls and long-sleeved jacket, coveralls, one- or two-piece chemical-splash suit, disposable chemical-resistant overalls)

- Coveralls *

- Gloves, outer, chemical-resistant

- Gloves, inner, chemical-resistant

- Boots, outer, chemical-resistant, steel toe and shank

- Boot covers, outer, chemical-resistant (disposable) *

- Hard hat

- Two-way radios

- Face shield

Level C

Level C PPE is indicated when the concentration(s) and type(s) of airborne substance(s) is known and the criteria for using air purifying respirators are met. The following constitute Level C equipment:

- Full-face or half-mask, air-purifying respirators

- Hooded chemical-resistant clothing (overalls, two-piece chemical-splash suit, disposable chemical-resistant overalls)

- Coveralls *

- Gloves, outer, chemical-resistant

- Gloves, inner, chemical-resistant

- Boots, outer, chemical-resistant, steel toe and shank

- Boot-covers, outer, chemical-resistant (disposable) *
- Hard hat
- Escape mask *
- Two-way radios
- Face shield *

Level D

Level D clothing is a work uniform affording minimal protection, used for the elimination of nuisance contamination only.

Level D PPE consists of the following gear:

- Coveralls
- Gloves *
- Boots/shoes, chemical-resistant, steel toe and shank
- Boots, outer, chemical-resistant (disposable) *
- Safety glasses or chemical-splash goggles
- Hard hat
- Escape mask *
- Face shield *

LIMITATIONS AND COMPATIBILITY OF CHEMICAL PROTECTIVE CLOTHING

As you may have learned in your initial HAZWOPER training, human skin is often considered the largest organ of the body, with a surface area of about 1.7 m². At this size, skin can absorb a wide range of toxic substances rapidly leading to acute or chronic injury.

Worker protection in hazardous materials emergency response focuses on respiratory and skin protection. For most severe chemical hazards, selection of respiratory protection is relatively simple. Maximal hazards generally require some form of self-contained breathing apparatus (SCBA) worn inside a totally encapsulating chemical protective suit (TECP suit). For Level B responses, the same SCBA equipment is worn with some other arrangement of chemical protective clothing (CPC).

While the selection of respiratory protection is somewhat simple, the selection of CPC is orders of magnitude more difficult. CPC is provided by dozens of manufacturers throughout the United States in a range of plastic and rubber materials. While some of these materials may be visually indistinguishable one from another, their performance, when challenged by a hazardous material, can be radically different. For example, a glove which is resistant to a broad range of solvents may rapidly dissolve on contact with sulfuric acid. Such dilemmas and the problems they create for incident commanders and emergency responders have led to numerous studies on the compatibility of chemical clothing.

A range of factors affects the performance of CPC in an emergency situation. CPC may be attacked by chemicals in three ways:

- **Permeation:** the process by which a chemical dissolves into or passes through a chemical protective material at a molecular level.

- **Penetration:** the bulk movement of a chemical through pores or flaws in a CPC garment. Penetration may occur through punctures, closure assemblies (such as zippers), or imperfect seams in a CPC garment.

- **Degradation:** the reduction of the chemical resistance or physical barrier provided by a CPC garment. Degradation may occur as a result of abrasion, physical wear, chemical exposure, or exposure to the elements.

CPC materials have varying rates of resistance to attack in each of these modes. In commercial literature, the resistance of CPC to attack is generally reported in two ways:

- **Breakthrough time:** the time elapsed from the application of chemical substance to the outside of a CP material to its detection on the inside.

- **Permeation rate:** the amount of a chemical which passed through a given area of clothing per unit time (such as $mg/m^2/hr$). Permeation is controlled by these four factors:

- diffusion coefficient of the permeating chemical in the material (lower diffusion, slower permeation)

- the difference (gradient) in chemical concentrations between the inside and outside surfaces of the material (higher concentration, faster permeation)

- the thickness of the material (thicker material, slower permeation)

- the area of the material making contact with the chemical (greater area, greater permeation)

Four basic principles should be used to guide the incident commander in the selection of chemical protective clothing:

- **No panacea:** No single chemical protective material provides protection against all substances. There is no universal chemical protective clothing.

- **No impermeability:** Since permeation is a physical phenomenon which occurs on a molecular level, there are no materials which are truly impermeable. While certain metals may have infinitesimally small permeation rates, there is still some transport at the molecular level. As metals are not feasible as CPC, we must work with materials, such as elastic-coat fabric, which have a higher permeation rate.

- **Specific details:** The better able the incident commander is to judge the exposure to a specific chemical hazard, the easier it is to purchase the ideal chemical protective clothing. CPC which offers the longest breakthrough time and the lowest permeation rate is the best.

- **Not for firefighting:** CPC does not offer thermal protection and may be flammable. Special protective clothing (such as aluminum oversuits) may be purchased for workers who may be exposed to fire in a hazardous materials response.

Finally, selection of CPC is complicated by the limited information provided by manufacturers. Much of this information focuses on a limited number of chemicals, most likely those for which the clothing has the best or the worst performances. Many other chemicals are not covered in manufacturer's data. CPC users should consult detailed tables of compatibility for assistance.

CARE AND FEEDING OF PERSONAL PROTECTIVE EQUIPMENT

The care of PPE can be broken down into six stages:

- Pre-use inspection
- Donning
- In-use monitoring
- Doffing
- Storage
- Maintenance

Pre-Use Inspection:

The following general inspection procedure is useful for all CPC items:

- Determine whether the material is correct for the problem at hand.
- Make a visual inspection for imperfect seams, nonuniform coating, tears, and malfunctioning closures.
- Hold the garment up to the light and check for holes.
- Flex the garment and check for cracks or other signs of deterioration.
- If the garment has been used previously, inspect it inside and out for signs of chemical degradation, including discoloration, swelling, or stiffness.

For totally encapsulating chemical protective suits, the following procedures should be observed:

- Check the operation of the pressure relief valves.
- Inspect the fit of the wrist, ankle, and neck seals.
- Check the face shield for cracks, crazing, or other fogginess.
- Ensure that the suit has received periodic pressure testing or in-use testing as described in Appendix A of 29 CFR 1910.120.

Donning of PPE:

Responders should establish a routine for the donning of PPE. All equipment should be carefully inspected prior to donning and immediately upon doffing. A field fit-check for the respiratory protection equipment should always be performed.

In-Use Monitoring of PPE:

Workers should develop a sense of how the PPE performs and should not continue to work when they feel that this performance has been impaired. Any sort of degradation of materials should be noted and appropriate action be taken.

Workers should be alert for any perception of odor, sense of burning in the eyes, or skin irritation while using the suit. Such sensations may be the sensory response from penetration of the suit.

While exposure to chemicals is the greatest hazard of PPE failure, other components of failure such as increased heat stress from faulty relief valves, failure of communications equipment, or decreased mobility should all be dealt with as serious.

Doffing of PPE:

As with donning, doffing should follow an established routine with inspection of every piece of equipment being removed.

Storage of PPE:

As HAZMAT clothing is often not used on a regular basis, its storage for eventual use in an emergency is of great importance. PPE should be stored away from potentially damaging sources such as dust, moisture, sunlight, damaging chemicals, temperature extremes, or physical abrasion.

In addition, after the material is used, it remains potentially contaminated, despite the best efforts at decontamination. Any potentially contaminated clothing should be stored in an area separate from street clothing. The storage area should be well-ventilated with good air flow around each garment.

For unused materials, different ensembles of CPC should be stored separately to avoid mixing CPC types. CPC should be stored in accordance with manufacturer's recommendations.

Maintenance of PPE:

Maintenance can generally be considered at three levels: user, shop, and supplier (or specialized). Depending on the needs of the facility, a maintenance program may be set up to provide competent repair at each of these levels.

LIMITATIONS OF THE HUMAN BODY IN PERSONAL PROTECTIVE CLOTHING

The aim of personal protective clothing, especially the totally encapsulating chemical protective (TECP) suit, is to keep hazardous chemicals away from the body. In so doing, these suits trap body heat and moisture against the skin. While suits are outfitted with relief valves and some air circulation units, they nevertheless constrain the movements of the wearer and place limits upon the ability to work.

HEAT STRESS

Heat stress is the total heat load which the body experiences from both external heat and from exertion (metabolic heat).

Heat is transferred away from the body in two ways, convection and conduction. Convection is the transfer of heat through a liquid or gas, while conduction is the transfer of heat through a solid. In humans, cooling of the body is primarily accomplished by means of the evaporative cooling of blood vessels close to the skin through sweating. As the body temperature increases, we sweat more profusely and that sweat evaporates. The process of evaporation cools the blood flowing in vessels close to the skin. Employees working in PPE will find evaporative cooling to be less efficient when little or no skin is exposed to the air.

The research detailed in the FEMA Publication, *Field Evaluation of Chemical Protective Suits* (FA-108, 9/91), describes the temperature extremes to which a body can be subjected working in a TECP suit. For example, after only a brief time under hot and wet conditions (92 degrees, 90 percent humidity), body temperature climbs to 100.4 °F and the heart rate to 152 beats/minute.

One of the key points of the medical surveillance plan developed by a facility with a duly constituted HAZMAT team should be a clear understanding of how employees will be instructed concerning heat stress and how heat stress will be monitored in employees responding to emergencies.

Employees should be instructed by a physician or other suitable medical practitioner to monitor their vital signs, especially heart rate. Workers should be limited to a certain percentage of the "maximum attainable heart rate," generally considered to be 220 beats per minute. Some authorities have developed a maximum heart-rate formula based on 70 percent of the maximum at an age-adjusted rate. The formula for age-adjusted maximum heart rate is

$$(0.7)(220\text{-age})$$

For example, a 35-year-old responder would have a maximum age-adjusted heart rate of 129.5 beats per minute calculated as follows:

$$(0.7)(220\text{-}35)$$

$$(0.7)(185) = 129.5 \text{ beats per minute}$$

In severe emergency response situations, employees should be required to take adequate replacement fluids during any breaks. Needless to say, care should be taken to avoid employee contamination during these breaks. Some cooling devices for hazardous materials suits do exist. However, such devices are not recommended for use under 95°F. These units are bulky and at lower temperatures actually increase the stress rate above that caused by increased temperature.

The final step in the protection of workers from stress should be the adequate and accurate education of workers in the symptoms and effects of heat stress. Prevention of further accidents in emergency response is worth more than a pound of cure.

HEAT-RELATED ILLNESSES

Heat rash: Heat rash, or prickly heat, results from hot and humid conditions. Rashes generally develop on body areas exposed to constant dampness, especially feet, underarms, and the crotch area. Although heat rash is not a serious condition, it is uncomfortable and reduces worker effectiveness.

Heat syncope: Heat syncope is the sensation of lightheadedness or fainting while standing in hot conditions. This results from blood pooling in the lower extremeties resulting in a momentary depletion of oxygen to the brain.

Heat exhaustion: Heat exhaustion is caused by dehydration due to sweating. Symptoms of heat exhaustion include dizziness, headache, nausea, confusion (disorientation), and fainting. A heat exhaustion victim will appear drenched with sweat and the skin will be cool and clammy.

Heat cramps: Heat cramps result from the loss of body electrolytes (essential elements) through sweating. Painful cramps in the abdomen and legs will result after exertion. Although not serious, heat cramps are a good indicator of heat stress.

Heat stroke: Heat stroke is the most serious of the heat-related illnesses. Individuals suffering heat stroke are in immediate danger of death.

COLD STRESS

Although for most in-plant emergencies, heat stress will be the dominant concern, cold extremes also present hazards to the emergency responder.

As most of us have experienced, shivering is the body's first response to cold. Shivering is caused by rapid and involuntary muscle contractions. These contractions produce heat with which the body attempts to compensate for our cold surroundings. Severe shivering may occur when our core temperature falls to below 95°F.

COLD-RELATED ILLNESSES AND INJURIES

Hypothermia: Hypothermia is overall cooling of the body core. Extremes of temperature are not necessary to produce hypothermic conditions. Wind chill and wet clothing may cause sufficient chilling. Hypothermia symptoms include a sensation of cold, uncontrollable shivering, pain in the extremities, and disorientation. Hypothermia may lead to unconsciousness and death.

Frostbite: Frostbite is, quite simply, the freezing of skin or tissue at the extremities. Freezing of the skin may be a minor injury, but deep frostbite may lead to the loss of fingers, toes, or other body parts. A signs of frostbite is white or grayish-yellow skin.

PHYSICAL LIMITATIONS OF PPE

As anyone who has used either personal protective equipment or scuba diving gear knows, the added burdens of the extra equipment range far from the physiological factors. In a word, PPE is bulky. PPE can limit vision, reduce mobility and maneuverability, and its use can induce stress and fear in employees who have a tendency towards claustrophobia.

The incident commander has the responsibility of ensuring that the physical and mental well-being of each employee serving in a HAZMAT role is suited to that role. The incident commander should not allow employees to respond to incidents if there is any doubt about their ability to perform.

5

PERFORMING A FACILITY EMERGENCY RESPONSE AUDIT: UNDERSTANDING FACILITY HAZARDS

INTRODUCTION

It has been said that knowledge is power. Certainly in terms of a safe and productive workplace, knowledge of workplace hazards is the power which prevents injuries, emergencies, and property damage. Just as traditional safety concerns mandate the use of a "job safety analysis," hazardous materials response professionals emphasize a "workplace safety analysis" which alerts workers to the range of problems which may exist for them in the workplace.

Completing an analysis of the workplace requires an understanding of the range of workplace hazards and the proper response to their occurrence. Much of the information needed to complete an analysis is provided to employees during their HAZWOPER First Responder: Awareness Level training.

The basic classes of hazard to be considered are the following:

- Oxygen deficiency
- Fire and explosion
- Safety hazards
- Electrical hazards
- Heat stress
- Cold exposure

27

- Noise
- Ionizing radiation
- Biological hazards
- Confined space entry
- Chemical exposure

Naturally, the emphasis of a HAZWOPER program falls on the last category: chemical exposures. However, emergency responders and incident commanders must be familiar with the broader range of hazards which may present themselves during an emergency.

For example, fires and explosions present a certain set of hazards to the emergency responder. Firemen responding to a fire will be well aware of the hazards they face from the fire alone. However, the firefighters may also face hazards from downed electrical lines or falling objects. By the same token, the emergency responder must take note of his surroundings before beginning an emergency response. Care must be taken to reduce these hazards and increase the chance for a trouble-free emergency response.

PERFORMING A PROCESS HAZARDS ANALYSIS (PHA)

Since the development of the OSHA Process Safety Management Standard [29 CFR 1910.119] in 1992, much attention has been paid to the concept of a process hazards analysis (PHA). This analysis is undertaken to determine the likely modes of failure of certain operating equipment containing hazardous chemicals.

While the Process Safety Management Standard mandates a PHA for only a select few processes, the concept can be expanded to cover processes which may not contain highly toxic substances, but whose contents, if released, would require response by a HAZMAT team.

A PHA consists of three basic steps:

Step 1: Hazard Identification

This process should occur in a systematic manner, on a hazard-by-hazard basis. During this procedure, it is important to keep in mind the following points:

1. The objectives are to prepare a complete list of all significant process hazards for control and attention by operational safety standards or administrative or engineering controls.

2. Although not all the team members may be experts in the technology, the team can determine the adequacy of the technology and the possible need for further technology development.

3. The purpose is to identify inherent hazards as well as deficiencies or concerns of the moment.

4. "What-if" thinking in a disciplined manner is used to search out all significant hazards inherent in the process.

5. Many hazards will have been previously identified in safety/health reports, accident/incident reports, pre-startup safety reviews, or operational safety/ health surveys.

6. The following areas should receive special attention:

    A. startup

    B. shutdowns (emergency and planned)

    C. standby operation

    D. upset conditions

    E. utilities failure

    F. multiple-jeopardy situations

Step 2: Hazard Qualification

Once the hazards in the facility have been identified, they must be quantified. Traditionally, risk has been viewed by the equation:

$$\text{Level of Risk} = \text{Severity of Risk} \times \text{Probability}$$

Hence, those analyzing the risks of an operation must be willing to quantify the severity of this risk based on the likelihood of damage to human health, property, and the environment from a given release. Some format must be developed to attach a number to that risk. For example, severity might be measured as follows:

*   Slight to none

*   Low (light injury/small property damage/minor environmental effect)

*   High (severe injury/property damage/some environmental damage)

*   Catastrophic (fatality/major property damage/major environmental damage)

By the same token, the probability of occurrence must be measured by some objective measure. Some of these probabilities can be obtained through manufacturer's specifications (mean time between failure: MTBF) or through operating history (e.g., a process engineer's knowledge that this valve has failed three times in the past seven years). A probability index can be developed equivalent to the risk index:

1. Very low ($< 10^{-4}$ per year, or less than one occurrence per 10,000 or more years)

2. Low (10^{-4} to 10^{-2} per year, or one occurrence per 100 to 10,000 years)

3. Medium (10^{-2} to 1 per year, or one occurrence per 1 to 1000 years)

4. High (one occurrence or more per year)

Step 3: Consequence Analysis

Finally, the risk analysis must take into account other factors in evaluating facility risks. While death, property loss, and environmental damage are the three big possibilities, other factors must be considered. Some possible consequences of a hazardous materials emergency:

*   Facility shutdown

*   Other business disruption

- Loss of public image
- Difficulty in re-building plant or unit

Brainstorming within the PHA team can determine other hazards which might apply to a specific facility.

PERFORMING A WORKPLACE HAZARD ANALYSIS

A workplace hazard analysis is the first step of the facility emergency response audit. It will include some of the features of the HAZOP analysis, but less time will be devoted to the engineering aspects of the processes and more to the potential indicators of irregularity.

As in many crime novels, it is not the ordinary which catches the eye of the workplace detective. He must be alert for the irregular. The process which is not functioning properly—for example, the drum which is improperly stacked—might be the indicator of a workplace hazard which could result in a hazardous materials release.

The basic workplace hazard analysis can be broken down into seven steps:

1. Determine location
2. Examine container/vessel/piping condition
3. Determine physical state
4. Determine dispersion pathways
5. Look for exposure indications
6. Look for safety hazards
7. Label and store

Step 1: Determine Location

It is important for the employee (or the incident commander) to have a good "feel" for the locations of hazardous materials, hazardous processes, or hazardous operating equipment. The analyst should note the locations of any containers or other storage systems:

- Drums
- Pallets
- Above- and below-ground tanks
- Gas cylinders
- Waste containers

Step 2: Examine Container Condition

Once the sizes and types of containers in the workplace have been determined, the incident commander should ensure that all stored materials are in acceptable condition. In addition, incident commanders should ensure that all applicable environmental (local, state, and Federal) regulations concerning the safe storage of hazardous materials are followed. Such concerns might include the following: secondary containment, segregation of incompatibles, limited access, and signage. All storage containers, vessels, and pipes should be inspected for the following aspects of their physical condition:

- Rust or corrosion

- Leaks

- Bulges

- Types and quantities of materials

- Labeling

Incident commanders may wish to review Article 80 (hazardous materials storage) of the uniform fire code. Although the uniform code is not adopted in all jurisdictions, the code document provides a good review of current thinking about hazardous materials storage. The uniform fire code is available from the International Conference of Building Officials (www.ICBO.com), 5360 Workman Mill Road, Whittier, California 90601-2298.

Step 3: Determine Physical State of the Contents

Once the contents of containers have been determined, care should be taken to determine the physical state (gas, liquid, or solid) of the contents. Clearly, the state of the material has an effect on the response to a leak or spill of that material. In addition, employees working with a certain chemical should know the normal color and physical appearance of that material. Changes in appearance can often indicate changes which may place the employee at risk in using the chemical.

Materials should not be subjected to swings in climate (hot/cold, wet/dry) during storage. Such changes can cause deterioration and physical changes in materials. A good example of this is latex paints exposed to freezing temperatures. Such exposure separates the water and other components of the emulsion, and they cannot be mixed again.

Step 4: Determine Dispersion Pathways

An incident commander auditing the hazards of a facility should always pay close attention to the paths by which hazardous materials may leave the facility. Many older operating facilities have in-plant process drains or floor drains which may be tied into the sanitary sewer or the storm sewer. Outside of the facility, the incident commander should be aware of any open storm drains or any access off the property which will lead hazardous materials to surface water.

In addition, incident commanders should have a clear understanding of the weather conditions prevalent at their site. Provision should be made for an accurate indication of the wind direction at all times. The incident commander should know the direction and relative average speed of the prevailing winds from the facility. Facilities with potentially airborne toxics should have one or more windsocks mounted in highly visible areas.

Lastly, the facility incident commander should be familiar with the site's groundwater. In an ideal situation, the incident commander would know the direction and rate of flow of the groundwater away from the facility. Such measurements may be made when baseline groundwater samples are taken.

Step 5: Look for Exposure Indicators

The incident commander should be alert to any indicators of potential exposures. Such signs can range from a subtle sheen of oil on surface water to dead animals in the vicinity of the facility. Site inspections should include reviews of stained soils or damaged vegetation.

Step 6: Look for Safety Hazards

The incident commander should be part of the facility team which inspects for safety hazards. Restricted access heights, projections which might snag chemical protective clothing, and other such safety hazards should be noted. This information should be collected and given to the hazardous materials responders during their briefing prior to entry into the hot zone.

Step 7: Label and Store

All stored materials should be properly labeled. Regardless of whether the labels are those supplied by the manufacturer or labels placed on the materials at the facility, labels should be readily visible during the inspection.

Storage facilities should be inspected to ensure that no incompatibles are stored together and that no safety hazards exist in the storage area.

BASELINE AUDITING AND PERIODIC REVIEWS

Facility auditing for emergency response compliance establishes a baseline from which future compliance may be benchmarked. The establishment of workplace hazard analyses is one small part of a comprehensive audit program. Such a program might contain the following components:

- Written program review:

  a) Evaluate the 12 points of the HAZWOPER emergency response program

  b) Evaluate the written respirator program

  c) Evaluate the written evacuation plan

  d) Evaluate the written training program

- Record keeping review:

  a) Review training records

  b) Review preventive maintenance and inspection records

  c) Review corrective action records

- Training program review:

  a) Evaluate effectiveness and content of training program; and

  b) Evaluate appropriateness of training program.

- Evaluate facility hazards:

- a) Inspect chemical storage areas

- b) Inspect hazardous waste storage areas

- c) Evaluate workplace hazards

- Evaluate response team readiness:

- a) Inspect equipment and response materials

- b) Inventory spill control equipment

- c) Exercise emergency response team

OSHA has prepared a detailed compliance guideline for Process Safety Management [OSHA CPL 2-2.45] which includes an audit section for emergency response.

LOCAL AND STATE EMERGENCY RESPONSE PLANNING

INTRODUCTION

In December 1984, a cloud of highly toxic methyl isocyanate spewed from a chemical plant in Bhopal, India, blanketing the surrounding area in poison. The result: over 2,000 people dead and thousands more injured. Injuries included damaged lungs, reduced oxygen flow, severe headaches, and temporary blindness. Even today, poor health continues to afflict this community. The gravity of this tragedy opened the eyes of the world to the dangers of chemical accidents. Eight months later, a less toxic derivative of that chemical escaped from a West Virginia plant (from a plant owned by the very company involved in the Bhopal disaster), bringing these same concerns home to the United States.

Accidents can happen at any facility, given certain circumstances. In Bhopal, prevention equipment had been installed and a local evacuation plan developed. Unfortunately, the equipment was not in service, and the neighboring community was not aware of the plans. This lack of knowledge proved fatal.

Based on these facts, the U.S. Congress passed the Emergency Planning and Community Right-to Know Act (EPCRA), as Title III under the reauthorization and amendment to Superfund (the Superfund Amendments and Reauthorization Act; SARA).

EPCRA is divided into four major sections:

1. Emergency Planning (Sections 302 and 303)

2. Emergency Notification (Section 304)

3. Community Right-to-Know On Chemicals (Sections 311 and 312)

4. Toxic Chemical Release Inventory (Section 313)

EMERGENCY PLANNING REQUIREMENTS: GOVERNMENT

The Governor of each state was required to establish a State Emergency Response Commission (SERC) by **April 17, 1987**. The SERC then was responsible for establishing local emergency planning districts by **July 17, 1987**. Each district would then be controlled by a local emergency planning committee (LEPC) consisting of members of local emergency response groups, community groups, and facilities owners and operators. These LEPCs were to be formed by **August 16, 1987**.

Each LEPC must appoint a chairperson, establish rules for public notification of committee activities, and establish procedures for receiving and processing requests for public information, including the designation of an information coordinator.

By **October 17, 1988**, all LEPCs were required to have an emergency plan for submission to the SERC for review and revision. The plan must be reviewed at least annually or more often if needed. The plan must do the following:

- Identify facilities subject to requirements

- List methods and procedures in response to a release

- Appoint community and facility emergency coordinators

- Include procedures for notification of releases

- Outline methods for determining the occurrence of a release and the sites and populations affected

- Describe facilities within a community and their emergency equipment

- Implement evacuation plans and a schedule to practice them

- Set up training programs and alternate traffic routes to prepare for an emergency response

- Include methods and schedules for exercising the emergency plan

Each facility incident commander is responsible for coordinating programs with the LEPC. It is imperative that the incident commander ensure that the information provided to the LEPC concerning the facility is accurate and up-to-date. The facility incident commander should make sure that the LEPC has the most up-to-date documentation from the plant. Local emergency response plans range from complex (for many major cities) to ill-formed documents. Concerned and active facility ICs can provide valuable assistance in developing a local plan.

EMERGENCY PLANNING AND NOTIFICATION REQUIREMENTS: INDUSTRY

Section 302(c) of the law requires that any facility having extremely hazardous substances (EHS) present above the threshold planning quantity for each particular substance be subject to the emergency planning requirements. A comprehensive listing of these regulated substances ("Regulated Substances for RMP and PSM Rules") can be found in Appendix H.

The facility's initial responsibility is to notify the SERC of their status as covered under the emergency planning provisions of the act. This notification was to be made by **May 17, 1987**. If a facility starts using, producing, or storing an EHS above the designated threshold quantities, it must report to the SERC within 60 days of the chemical's presence at the site.

As spelled out in Section 303(d)(1) of the law, all industrial facilities subject to the Act must designate an emergency coordinator by **September 15, 1987**. In addition, owner and operators of the regulated community must cooperate with the local committees, and provide information on any changes to the facility. If the designated coordination has changed since this date, proper notification should be made to the LEPC.

Section 311 of the Act [40 CFR 370.21] requires the owner or operator of a facility subject to the emergency planning requirements of EPCRA to submit a Material Safety Data Sheet (MSDS) to the SERC, LEPC, and local fire department for each hazardous chemical present at the facility. The owner is subject to this requirement based on the following minimum thresholds:

- The owner or operator of a facility shall submit a Material Safety Data Sheet (MSDS) on or before **October 17, 1987** (or within three months after the facility becomes subject to these requirements) for all hazardous chemicals present at the facility at one time in amounts greater than 10,000 pounds and for all extremely hazardous substances present at the facility in an amount greater than or equal to 500 pounds (approximately 55 gallons) or the TPQ, whichever is lower.

Reporting requirements under Section 312 (Tier I and Tier II Emergency and Chemical Inventory) and Section 313 (Toxic Chemical Release Inventory, Form R) had initial submittal dates of **March 1, 1988** and **July 1, 1988**, respectively, and yearly thereafter. These inventories detail the hazardous substances storage and usage within manufacturing facilities.

PROCEDURES FOR SARA TITLE III COMPLIANCE

1. Using chemical inventory information, determine the presence of Extremely Hazardous Substances (EHS) at the facility above the Threshold Planning Quantity (TPQ). If your facility has EHS above the TPQ, then you are subject to the emergency planning requirements of SARA.

2. Verify the Reportable Quantity for EHS at the facility.

3. Notify the State Emergency Response Commission (SERC) that the facility is subject to the emergency planning requirements of SARA.

4. Assign an Emergency Coordinator.

5. Prepare emergency planning procedures in the event of a release of an EHS in an amount which exceeds the RQ.

6. Submit copies of MSDS for *all* chemicals, or a list of chemicals (grouped by health and physical hazard), for which a MSDS is required, to the SERC, LEPC, and fire department.

7. Notify the SERC within 60 days if a EHS becomes present at your facility in quantities which exceed the TPQ.

FEDERAL EMERGENCY RESPONSE

A key element of Federal support to local responders during hazardous materials incidents is a response by U.S. Coast Guard (USCG) or EPA On-Scene Coordinators (OSCs). The OSC is the federal official predesignated to coordinate and direct Federal responses and removals under the National Contingency Plan (NCP). These OSCs are assisted by federal Regional Response Teams (RRTs) that are available to provide advice and support to the OSC and, through the OSC, to local responders.

Federal responses may be triggered by a report to the National Response Center (NRC), operated by the Coast Guard. Provisions of the federal Water Pollution Control Act (Clean Water Act), CERCLA, and various other federal laws require persons responsible for a discharge or release to notify the NRC immediately. The NRC duty officer promptly relays each report to the Coast Guard or EPA OSC, depending on the location of the incident. Based on this initial report and any other information that can be obtained, the OSC makes a preliminary assessment of the need for a federal response.

This activity may or may not require the OSC to go to the scene of an incident. If an on-scene response is required, the OSC will go to the response and monitor the response of the responsible party or state or local government. If the responsible party is unknown or not taking appropriate action, or the response is beyond the capability of state and local governments, the OSC may initiate federal actions. The Coast Guard has OSCs at 48 locations (zones) in 10 districts, and the EPA has OSCs in its 10 regional offices and in certain EPA field offices.

REGIONAL RESPONSE TEAMS

Regional Response Teams are composed of representatives from federal agencies and a representative from each state within a federal region. During a response to a major hazardous materials incident involving transportation or a fixed facility, the OSC may request that the RRT be convened to provide advice or recommendations on specific issues requiring resolution.

An enhanced RRT role in preparedness activities includes assistance for local community planning efforts. Local emergency plans should be cooordinated with any federal regional contingency plans and OSC contingency plans prepared in compliance with the NCP.

THE NATIONAL RESPONSE TEAM

The U.S. EPA and U.S. Coast Guard have developed the National Response Team (NRT) as a national repository for both technological and educational materials related to emergency response. The NRT is a repository for publications (books, software, and videos) related to emergency response. The NRT can be reached at 201-321-6740.

7

SPILL AND RELEASE REPORTING UNDER FEDERAL REGULATIONS

INTRODUCTION

Spill reporting is controlled by three regulations in the United States: the Comprehensive Environmental Response, Compensation, and Liability Act (CERCLA), the Department of Transportation (DOT) Hazardous Materials Transportation Act, and the Clean Water Act. This chapter examines the reporting requirements under these three laws.

SPILL REPORTING UNDER CERCLA AND THE CLEAN WATER ACT

While much attention is paid to the provisions of CERCLA and its application to Superfund sites, CERCLA is extremely important for producers, transporters, and users of hazardous substances.

The National Oil and Hazardous Substances Pollution Contingency Plan or National Contingency Plan (NCP) was originally outlined in Section 311 of the Clean Water Act. The plan originated to give the Coast Guard primary responsibility in responding to oil spills (40 CFR 110-114) and has expanded to focus on discharges of hazardous wastes and the interaction between the CWA and CERCLA (Superfund) (40 CFR 116-117).

Primarily, the NCP requires the immediate reporting of any discharges to the National Response Center (NRC) 800-424-8802 (or 202-426-2675 in Washington, DC) and outlines the quantities and severity of spills.

The NCP outlines the process for developing an Oil Removal Contingency Plan for each facility at risk of a discharge. Oil and hazardous substance spills are classified as follows:

Minor: less than 1,000 gallons of oil in the inland waters, less than 10,000 gallons of oil in the coastal waters, or less than a reportable quantity (RQ) of a hazardous substance.

Medium: 1,000 to 10,000 gallons of oil in the inland waters, 10,000 to 100,000 gallons of oil in the coastal waters, or a hazardous substance equal to or greater than a reportable quantity.

Major: more than 10,000 gallons of oil in the inland waters, more than 100,000 gallons of oil in the coastal waters, or a discharge of a hazardous substance that substantially threatens public health or welfare or results in critical public concern.

The NCP also requires facilities which have discharged oil or could discharge oil in the future to develop a comprehensive spill prevention control/countermeasure (SPCC) plan. Such plans must be developed and reviewed by a registered professional engineer. The SPCC plan should outline possible pathways for oil discharges from facilities and detail control measures which could be taken to prevent the spill from damaging the environment. A summary of SPCC contents can be found in Appendix D.

INFORMATION TO BE REPORTED

1. Identity of person making report to NRC
   a. Job title
   b. Location(s) where s/he can be reached
   c. Date and time

2. Identity and amount of substance released

3. Geographical location of release

4. Reason for classifying material as a hazardous substance

5. The reportable quantity of each substance involved

6. Circumstances of release (to ground, water, air)

7. Date, time and duration of release

8. Remedial actions taken to control, and/or mitigate the effects of the release

9. Regulatory agencies to which you have given notice

DETERMINING IF NOTIFICATION IS REQUIRED

CERCLA requires any person in charge of a facility to notify the National Response Center immediately upon discovery of any release from that facility of a hazardous substance in a quantity equal to or greater than the "reportable quantities" listed by EPA.

To decide if notification is necessary, you must determine the following:

1. Whether there has been an unpermitted release

2. Whether the substance released was hazardous as defined by CERCLA

3. Whether the amount released was large enough to be a "reportable quantity"

The NRC database (Emergency Response Notification System, ERNS) is available to the public and can be searched through a Freedom of Information Act (FOIA) request. A summary of the database contents is included as Appendix I.

UNPERMITTED RELEASES

A "release" is any spilling, leaking, pumping, pouring, emitting, emptying, discharging, injecting, escaping, leaching, dumping, or disposing of a substance into the environment. This includes releases onto the land, into the land, into groundwater, surface water, or into the ambient air. Examples include tank overflow, pond overflow, a container overturned in a yard, ruptured disk blowout, safety valve release, scrubber failure or bypass, fire, pipeline break, or an underground pipe leak.

There are six situations in which a release does *not* trigger notification requirements under CERCLA:

1. Any release that only exposes persons in a workplace. This would include a spill or release which occurs in a totally enclosed building, and does not leave the building.

2. Emissions from the engine exhaust of motor vehicles, rolling stock, aircraft, vessels, or pipeline pumping station engines.

3. Releases of source, by-product, or special nuclear material from a nuclear incident.

4. The normal application of fertilizers or pesticides.

5. Releases made under a federal permit. This includes releases that are covered by final permits under the following regulations:

   a. Clean Water Act

   b. Clean Air Act

   c. Atomic Energy Act

   d. Marine Protection, Research, and Sanctuaries Act of 1972 (ocean dumping)

   e. Safe Drinking Water Act (underground injection)

   f. Resource Conservation and Recovery Act (RCRA)

6. Continuous releases, if they are stable in rate and quantity and at least one notification is provided annually to establish the continuity and quantity of the release.

Once again, prompt and correct reporting under CERCLA is important. Enforcement of CERCLA's release reporting requirements has increased immensely during recent years. The EPA has cited numerous facilities for delays in reporting, now demanding reports within 15 minutes of the discovery of a release.

CERCLA HAZARDOUS SUBSTANCES

CERCLA identifies hazardous substances by incorporating lists of hazardous and toxic substances and hazardous wastes established by EPA under other statutes. This list is set forth in Table 302.4 (40 CFR 302.4).

In addition to this specific list, hazardous substances, for the purposes of spill reporting, include any hazardous waste subject to regulation by EPA under RCRA because it has one of the characteristics of a hazardous waste: corrosivity, reactivity, ignitability, or toxicity (as defined by the toxicity characteristic leaching procedure).

REPORTABLE QUANTITIES (RQ)

Five different reportable quantities (RQ) are used for various hazardous substances regulated under CERCLA, each with the category designation shown in brackets: 1[X], 10[A], 100[B], 1000[C], and 5000[D] pounds. To determine which RQ to apply, you must first determine whether the substance released is listed in the Title III List of Lists. If it is, the RQ for that substance will be given in the last column of the table.

Releases of mixtures and solutions are subject to these notification requirements only where a hazardous substance which is a component of the mixture or solution is released in a quantity equal to or greater than its individually reportable quantity.

For unlisted (EPA Characteristic) hazardous wastes, the reportable quantity is 100 pounds.

SPILL REPORTING UNDER DOT REGULATIONS

The Hazardous Materials Transportation Act stipulates that whenever hazardous materials are unintentionally released during transport, loading, unloading, or storage, the incident must be reported. Certain releases require immediate telephone notification, while in other cases detailed written reports suffice.

REPORTING PROCEDURES

1. Notify the National Response Center:

   (800) 424-8802   Washington, DC: (202) 267-2675

2. Provide the following information:

   a. Name of notifier

   b. Phone number of notifier

   c. Name and address of carrier

   d. Date, time, and location of incident

   e. Extent of injuries

   f. Classification, name, and quantity of hazardous materials or waste involved

   g. Type of incident and whether a continuing danger to life exists at the scene

DOT NOTIFICATION REQUIREMENTS

DOT regulations (49 CFR 171.15) require immediate notification by telephone if any of the following occur as a direct result of a hazardous materials incident:

1. A person is killed.

2. A person receives injury requiring hospitalization.

3. Estimated carrier or other damage exceeds $50,000.

4. An evacuation of the general public occurs lasting one or more hours.

5. One or more major transportation arteries or facilities are closed or shut down for one hour or more.

6. The operational flight pattern or routine of an aircraft is altered.

7. Fire, breakage, spillage, or suspected radioactive contamination occurs involving shipment of radioactive material.

8. Fire, breakage spillage, or suspected contamination occurs involving shipment of etiologic (disease-causing) agents.

9. A situation exists of such a nature that, in the judgment of the carrier, it should be reported, even though it does not meet the above criteria; e.g., if the situation at the scene presents a continuing danger to life.

DOT WRITTEN NOTIFICATION

A carrier must submit a detailed written report of any accidental discharge of hazardous materials or waste. This provision of the HMTA applies regardless of the need for telephone notification.

Within 30 days of the incident, the carrier must file two copies of DOT Form F5800.1, the DOT Incident Report Form.

In cases where the spill involved a hazardous waste, a copy of the hazardous waste (RCRA) manifest must be attached. The report must include an estimate of the amount of waste removed from the scene, name and address of the facility to which it was taken, and the method of disposal.

LEAKING CONTAINERS

The DOT regulations are stringent about the handling of leaking storage containers. From a shipper's perspective, employees should never offer a leaking container for shipment. Any leaking containers should be replaced, repaired, or over-packed before transportation.

The directions for proper handling of an over-pack or salvage drum are contained in the DOT regulations at 49 CFR 193.3(c):

"Salvage drums: Packages of hazardous materials that are damaged, defective, or found leaking and hazardous materials that have spilled or leaked may be placed in a metal or plastic removable head salvage drum that is compatible with the lading and shipped for repackaging or disposal under the following conditions:

(1) The drum must be a UN 1A2, 1B2, 1N2, or 1H2 drum tested and marked at least for the Packing Group III performance standard for liquids with a specific gravity of 1.2 and a hydrostatic pressure test of 35 kPa (5 psig). Capacity of the drum may not exceed 450 L (119 gallons).

(2) Each drum shall be provided when necessary with sufficient cushioning and absorption material to prevent excessive movement of the damaged package and to eliminate the presence of any free liquid at the time the salvage drum is closed. All cushioning and absorbent material used in the drum must be compatible with the hazardous material.

(3) Each drum shall be marked with the proper shipping name of the material inside the defective packaging and the name and address of the consignee. In addition, the drum shall be marked 'Salvage Drum.'

(4) Each drum shall be labeled as prescribed for the respective material.

(5) The shipper shall prepare shipping papers in accordance with subpart C of Part 172 of this subchapter."

From the standpoint of the carrier, drivers often discover leaking containers at a customer's dock, while unloading other containers. In this case, if the material is for the customer, the customer may take custody of the material and overpack it, or they may prevent the carrier from off-loading the container.

If the driver is prevented from off-loading the container, the customer's employee's must not allow him to leave the premises with the leaker and return to the truck terminal. Regulations state that the leak must be repaired as soon after it is discovered as is safe and practical.

To the employee on the receiving dock, this means that the carrier should be directed to a safe portion of the yard or a nearby safe space. The carrier's personnel should then be contacted and allowed to overpack the leaking container.

If the driver, with the knowledge of the dock employees, transports a leaking container beyond the nearest safe location, the driver and the dock employees are equally liable for all penalties that might be incurred.

SPILL REPORTING WORKSHOP

INTRODUCTION

This chapter consists of a workshop which will be used by the incident commander to determine his comprehension of the range of spill reporting regulations which may be in effect at his facility.

As discussed in the previous chapter, spills at facilities are covered by a range of regulations including CERCLA, the Hazardous Materials Transportation Act, and the Clean Water Act. In addition, OSHA requires reporting certain accidents and incidents which involve worker safety.

Read each of the following examples carefully. The answer follows each example.

EXAMPLE #1: AMMONIA RELEASE

At 4:00 A.M., a relief valve fails within a facility's ammonia refrigeration system. The valve is stuck open for approximately five minutes before the system's automatic shutoffs kick in and close the failing line down.

As a result of this failure, approximately 600 pounds of anhydrous ammonia is released from the system. However, since the release occurs in the center of a 90-acre facility complex, the release disperses before it reaches the perimeter of the facility.

Facility maintenance men are dispatched the next morning to repair the faulty valve. No further ammonia is released during the incident.

What actions should have been taken during this release?

_____

_____

_____

Who was responsible for taking them?

_____

_____

_____

Explain your reasoning.

_____

_____

_____

ANSWER #1: AMMONIA RELEASE

The ammonia release exceeded the reportable quantity (RQ) of 100 pounds. As such, it should have been reported immediately to the National Response Center. By not doing so, the facility personnel who identified the leak have breached their responsibility under the emergency reporting requirements of CERCLA.

Some readers may argue that the release did not have any off-site consequences as it dispersed before leaving the company property. But since the release occurred at 4:00 A.M., it was impossible for anyone at the facility to state with certainty the overall dispersion pattern of the ammonia.

Notification should be made immediately upon discovery, even if that discovery comes several hours later and the hazard has dissipated. Prompt notification not only fulfills the legal requirements; it also provides protection for the facility from future claims resulting from damages connected with the incident.

EXAMPLE #2: SULFURIC ACID RELEASE

A 55-gallon drum of sulfuric acid is punctured by a material handler. The puncture is severe enough to result in the loss of the entire contents of the drum. The area of the spill is evacuated and the facility HAZMAT team is called to neutralize the spill.

The spill is neutralized to a pH of 6.2, and the neutralized pool of liquid is washed into a process sewer using facility fire hoses. The process sewer discharges, through a holding sump, to the municipal sewerage system.

What actions should have been taken during this release?

_____

_____

_____

_____

_____

Explain your reasoning.

_____

_____

_____

_____

_____

ANSWER #2: SULFURIC ACID RELEASE

The release of materials to a process sewer is often an area of significant concern. Hazardous materials may result in serious damage to the biological treatment system of the municipal wastewater treatment plant.

The HAZMAT team may have isolated the process sewer prior to neutralization. In this instance, the pH as tested was acceptable for the sanitary sewer.

Municipal authorities may require a facility to complete a hazardous materials control, or slug discharge plan to provide a plan for such discharges.

EXAMPLE #3: CANOLA OIL RELEASE

During the unloading of a rail car containing canola oil, a fitting on the valve controlling flow into the receiving vessel ruptures causing a release of more than 200 gallons of oil onto the loading lot of the facility.

Employees wash the spilled oil from the lot using a combination of hot water and steam. The oil/water mixture flows into a storm sewer.

What actions should be taken regarding this spill?

_____

_____

_____

_____

_____

Explain your reasoning.

_____

_____

_____

_____

_____

ANSWER #3: CANOLA OIL RELEASE

This is a minor, but reportable release of oil to the storm sewer. If the storm sewer combines with sanitary discharges at a municipal wastewater treatment plant, the spill should be reported to the plant operator.

On the other hand, most storm sewers discharge to tributaries or navigable waters of the United States and are therefore subject to the reporting provisions of the Clean Water Act (report to NRC).

Spills to storm sewers should be considered during the development of the Storm Water Pollution Prevention Plan (SWPPP) required for most industrial facilities.

9

DEVELOPMENT OF EMERGENCY RESPONSE EXERCISE SCENARIOS

INTRODUCTION

Lying around on the couch for a couple of weeks, the average person will put on a couple of pounds, muscles will weaken, and stamina will decrease. This deterioration will not be stopped by a one-day hike in the woods or a swim in the ocean. By the same token, a well-trained emergency response team will lose its edge if not properly exercised. A once-a-year, eight-hour refresher course will do little to satisfy the needs of the facility's team in case of an emergency.

Emergency responders should be subjected to drills on a regular basis. Unfortunately, unlike fire drills, hazardous materials drills will likely require significant disruption of the facility and may even require the use of resources from outside the community.

The National Response Team (NRT) and the Federal Emergency Management Agency (FEMA) have developed a strategy for emergency response exercises which uses three levels of exercise:

1. Tabletop
2. Functional
3. Full-scale

This chapter examines these three exercise strategies and the use of the Hazardous Materials Exercise Evaluation Methodology (HM-EEM) for the evaluation of exercises.

TABLETOP EXERCISES

Occasionally you will read a story of an amateur runner who decides to enter a marathon as a first road race. Often these runners fail to complete the course. The reason: they have underestimated the task which lies ahead of them. In the same way, incident commanders with a freshly educated group of responders may be tempted to push these responders to participate in a full-scale exercise. When the smoke clears (sometimes quite literally), the facility will find that these new responders have been poorly prepared for the task at hand.

To prevent this and to ensure readiness of employees, many incident commanders will work response teams through emergency scenarios outlined on paper in order to determine the "thought" processes. Tabletop exercises help determine how the facility participants would work their way through an emergency, given limited and structured information.

Tabletop exercises are ideal for training and refreshing hazardous materials teams during months when they cannot perform tasks outdoors. Designers of tabletop exercises must focus on specific objectives which should be met in the completion of the exercise. A post-exercise critique should be developed and assessed upon completion.

An example of a tabletop exercise from FEMA Publication 254 is outlined as Appendix J.

FUNCTIONAL EXERCISE

As attractive as a tabletop exercise sounds, it is difficult to patch a leaking ammonia refrigeration system or a leaking tank car on a tabletop. Some emergency response skills must be practiced on the subject equipment, in full gear, and with defined procedures. Such activities are the focus of the functional exercise.

Rather than dealing with a particular series of emergency events, the functional exercise may focus on a specific action, such as patching a 55-gallon drum, which would be completed by each of the HAZMAT team members. A scenario would be developed which would call for a certain type of release, and the responders would receive information similar to that available during an actual incident. Each buddy group would be required to complete the task at hand.

Each participant would be required to don the proper PPE, move to the release area, and carry out the appropriate repair. A functional exercise could be completed without disruption to the facility and without exercising every portion of the emergency response plan.

A typical functional exercise scenario is outlined in Appendix K.

FULL-SCALE EXERCISES

For the emergency responder, the full-scale exercise is game day. Regardless of the level of preparation, an emergency situation is difficult and stressful. As an exercise, the emergency may not have the same environmental or human health impact, but those impacts will be simulated and the responders will be judged on their ability to ameliorate them.

A full-scale exercise is just what it says, a soup-to-nuts process which examines all of the functions of the emergency response system. A full-scale exercise may be limited to the facility, restricting the use of outside services or appointing surrogates to emulate those services. In its full form, the exercise may include all of the services surrounding the community: police, fire, ambulance, and hospital. Local media will be contacted and informed of the on-going emergency, and all of the functions of the response will be exercised.

The complexity of the full-scale exercise is limited only by the imagination of the developers. Most exercises use props such as liquid spills or smoke/fire bombs to enhance the realism of the exercise.

A full-scale exercise would require the development and exercise of all functions of the emergency response including the following activities:

- Implementation of the incident command system and the facility emergency response plan

- Site control and development of hot, warm, and cold zones

- Activities of the liaison officer to coordinate off-site emergency services

- Function of the public information officer in providing access to the media

- Proper hazard and risk assessment

- Selection and use of proper PPE

- Implementation of communication systems

- Decontamination, contamination reduction corridors, and disposal of wastes

- Termination and critique of the exercise

An example of a full-scale exercise description from FEMA Publication 254 is included as Appendix L. Additional examples are provided in the National Response Team publication: *Developing a Hazardous Materials Exercise Program* (NRT-2, 9/90).

USE OF THE HAZARDOUS MATERIALS EXERCISE EVALUATION METHODOLOGY (HM-EEM)

The Federal Emergency Management Agency has developed the Hazardous Materials Exercise Evaluation Methodology (HM-EEM, 2/92) for use by evaluators of full-scale exercises, especially those involving elements of the local, state, and federal emergency response organizations.

The HM-EEM focuses on 16 basic exercise objectives:

1. Initial Notification of Response Agencies and Response Personnel

2. Direction and Control

3. Incident Assessment

4. Resource Management

5. Communications

6. Facilities, Equipment, and Displays

7. Alert and Notification of the Public

8. Emergency Information: Media

9. Protective Actions for the Public

10. Response Personnel Safety

11. Traffic and Access Control

12. Registration, Screening, and Decontamination of Public

13. Congregate Care

14. Emergency Medical Services

15. Containment and Cleanup

16. Incident Documentation and Investigation

Each of these objectives has clear-cut evaluation elements and questions to be provided to the evaluator. Use of the HM-EEM for a facility functional or restricted exercise could be useful in pointing up potential pitfalls prior to an actual emergency or a full-scale exercise.

CONDUCTING THE EXERCISE

In any exercise, the level of alert and the amount of information provided to responders may vary. For the most part, complete surprise is impossible to achieve. Coordination with local and state officials is necessary to ensure that all facets of the response are exercised adequately. In addition, public safety agencies have to be given the option of calling the emergency in face of situations which would make it impossible for them to handle the emergency.

The most sophisticated exercises include the development of scenarios in secret without the participation of the facility and the facility incident commander. The resulting exercise serves the purpose of giving a full-scale alert to the facility and allows participants to interact with the local authorities who have been properly prepared.

Observers (evaluators) of the response can be organized from within the company, or they can be brought in from outside as consultants or through outside public agencies.

The exercise should be viewed as the "final exam" for a HAZMAT team, and care should be taken to be ready for that exam at all times.

10

DECONTAMINATION PROGRAMS

INTRODUCTION

Decontamination is the process of removing or neutralizing contaminants that have adhered to clothing and equipment used in hazardous waste emergency operations. The proper implementation of decontamination plans is crucial to the health and safety of workers at these sites. While the effects of decontamination on the workers in the hot zone is clear, an effective decontamination program will have three important side effects:

- Protection of all site workers by restricting the passage of contaminants into clean areas

- Prevention of possible reactions from inadvertent mixing of incompatible chemicals

- Protection of the community from contamination arising from transportation of contaminants from the incident site

Much contamination can be reduced or avoided by proper work practices. However, the overall gains from these practices are limited. Employees who are actively involved in the clean-up of hazardous materials will contaminate their equipment, gloves, and other clothing. As a result, the incident commander must develop a decontamination plan which will minimize the transport of contaminants during removal of protective equipment, cleaning of tools, and storage of equipment.

A decontamination program, focusing on the physical removal or chemical neutralization of the contaminants, must be developed for each compound likely to be released.

COMPONENTS OF A DECONTAMINATION PLAN

The decontamination plan for a facility should be a basic part of the facility's emergency response plan. The plan should focus on a "worst-case" scenario; that is, it should be prepared to deal with decontamination of the maximum levels of contaminants.

At a minimum the plan should do the following:

- Establish the number and layout of decontamination stations
- Determine the amount and types of decontamination equipment needed
- Detail the decontamination methods to be used at the facility
- Establish procedures to prevent the contamination of the cold zone
- Establish doffing procedures which are designed to minimize employee contact with contaminants during removal of PPE
- Establish methods for disposing of contaminated clothing

The incident commander should be prepared to alter (especially, to scale back) the original decontamination plan when the extent of the contamination is known.

METHODS OF DECONTAMINATION

Methods for decontaminating buildings, structures, and equipment can be grouped according to the mode of action. The methods include physical/mechanical, chemical, and thermal means, as well as solvent extraction. The following is a partial listing of methods:

- Asbestos abatement
- Absorption
- Bleaching
- Demolition
- Dismantling
- Dusting/vacuuming/wiping
- Encapsulation
- Grit blasting
- Hydroblasting/water washing
- Microbial degradation
- Painting, crating
- Scarification
- Solvent washing
- Steam cleaning
- Vapor-phase solvent extraction

Methods for decontaminating personal protective clothing and equipment may involve several different stages of washing or rinsing in various decontamination solutions or water. Some outer protective clothing may need to be disposed of, along with other cleanup wastes. Standard Operating Safety Guides have been published by the EPA which describe the various techniques used by emergency response personnel in decontaminating protective equipment.

PHYSICAL METHODS

Physical decontamination physically separates a chemical from the surface to which it adheres. Scrubbing is the most common method of physical decontamination used on PPE. Since physical decontamination uses no chemicals, it does not create secondary reactions which may further damage the equipment. It is a method which is inexpensive and easily applied. Finally, it is highly effective against a broad range of chemicals.

While physical decontamination has many advantages, the abrasion it uses does not remove all traces of most compounds and may cause more harm than good.

CHEMICAL METHODS

Chemical solutions can be produced which will interact with many substances to neutralize them and remove them from PPE. Chemical solutions have an advantage in that they can completely remove specific contaminants. On the other hand, chemical methods may cause subsidiary problems as they may mix and result in a toxic mixture. Disposal of such mixtures should be considered before selecting chemical decontamination methods.

ASSESSING THE LEVEL OF DECONTAMINATION NEEDED

The level of decontamination needed is dependent on the type and amount of contaminant present on the employee. Highly toxic substances may require additional levels of decontamination (more stations) to effectively complete decontamination.

The overall amount of contamination will be dependent on the level of physical contact the responder has had with the material during the response. A responder who must cross a puddle of solvent to reach a leaking valve or pipe will be exposed to greater quantities of material than a responder who has been called to clean-up a five gallon container of solvent.

SETUP AND MANAGEMENT OF A DECONTAMINATION AREA

Several good resources are available which detail the setup and management of decontamination areas. We have included two such documents as appendices to this chapter. Appendix M of this book is a reprint of Appendix D of the *NIOSH/OSHA/USCG/EPA Occupational Safety and Health Guidance Manual for Hazardous Waste Site Activities* (1985). Appendix N is a reprint of decontamination guidelines prepared by the Canadian Association of Fire Chiefs, entitled *Guidelines for Decontamination of Fire Fighters and Their Equipment Following Hazardous Materials Incidents*.

The decontamination area is the most important section of the hazardous materials emergency response area. No employee should be allowed to enter the response area (hot zone) until the decontamination area is established. If an employee must exit the area due to an emergency, such as a medical condition, the decontamination zone must be ready to receive him.

The decontamination area should have access to water and appropriate control equipment. The area should be set up separately from the area where tools, equipment, and vehicles are being cleaned.

A "contamination reduction corridor" (CRC) should be created by clearly demarcating the process lines where responders will be decontaminated. The CRC should be used to restrict access to the hot zone to those wearing appropriate PPE.

Appendices M and M outline the various sorts of process lines which are recommended depending on the nature of the contamination present.

The incident commander has responsibility for the management of the decontamination area. The incident commander should ensure that the goal of decontamination—to eliminate the transfer of contaminants from exposed equipment and clothing to any other surfaces—is met.

The incident commander should ensure that all personnel working in the decontamination area are properly protected and that their job responsibilities are clear.

11

MEDICAL SURVEILLANCE PROGRAMS

INTRODUCTION

One of the greatest responsibilities for the facility incident commander is the development and maintenance of a medical surveillance program. To some extent, it is the most difficult component of the job, made more difficult because it covers topics foreign to most facility personnel.

Medical surveillance programs include the development of health histories (physical exams, written histories) for members of the HAZMAT team and provide follow-up for individuals exposed to hazardous materials after an incident.

MEDICAL SURVEILLANCE REQUIREMENTS

HAZWOPER requires employees in four classes to be included in medical surveillance programs:

- workers whose exposures may exceed OSHA Permissible Exposure Limits (PELs) or other published exposure limits

- workers who wear respirators for 30 or more days per year

- members of HAZMAT teams

- workers who become ill or injured as a result of working at a HAZWOPER site or on a HAZMAT team

Employees within these classifications will be subject to one or more physical exams. For the purposes of this text, we will focus on the exams which would be encountered by a worker involved with a HAZMAT team. Such an employee would have the following exams:

- **baseline exam:** an exam completed prior to involvement in the HAZMAT team, usually the most comprehensive exam completed by the team member

- **annual exam:** the medical exam given at least every twelve (12) months to any HAZMAT team member, unless the attending physician determines that an interval not exceeding two years is more appropriate

- **exposure incident exam:** an exam given as soon as possible after a worker becomes ill or is injured working as a member of a HAZMAT team

- **termination exam:** an exam given when an employee stops working for the employer or is reassigned to new work that does not include the HAZMAT team

HAZWOPER-related exams should be provided by a licensed physician versed in occupational medicine. Of specific interest to the incident commander is the employee's fitness to work in self-contained breathing apparatus (SCBA) or in an air-purifying respirator. Specific requirements for all exams are provided in detail in the Joint Agency Hazmat Manual entitled *Occupational Safety and Health Guidance Manual for Hazardous Waste Site Activities* (NIOSH/OSHA/U.S. Coast Guard/USEPA, October 1985, NIOSH Publication No. 85-115).

The HAZWOPER standard requires that the physician providing the medical surveillance exam be provided with the following:

- a copy of the HAZWOPER rule [29 CFR 1910.120]

- a description of the worker's job, including any possible exposure levels to hazardous materials, heat stress, or other hazards associated with the HAZMAT team

- the personal protective equipment (PPE) needed for the job, including the type of respirator worn

- information from previous medical exams that the physician may not have on file which might relate to evaluating the employee's past and current health status

Once the physician completes the examination, the physician needs to complete a written opinion provided both to the worker and the employer. The written statement from the physician is designed to provide the employer the following information:

- any medical reason the worker couldn't safely perform his job on the HAZMAT team

- limitations the physician wants to place on the worker to protect his health on the job

- the results of the medical exams or tests (if requested by the worker)

- evidence of notification of the worker by the physician of the results of the medical exam

DOCUMENTING AND TRACKING MEDICAL SURVEILLANCE

In addition to all of the requirements for training documentation, facility incident commanders must develop a strategy for documenting and tracking the medical surveillance program. As with the training programs, the easiest method is to develop a tracking sheet or spreadsheet with the dates of initial exams and follow-up exams corresponding to the employee's activities and status. For employees with different job assignments, medical surveillance requirements may be of differing magnitudes.

Specific requirements under HAZWOPER require that the following documentation be kept for thirty years beyond employment:

- the worker's name and Social Security number
- the physician's written opinion of the worker's ability to perform the job safely
- a description of any limitations recommended for the worker by the physician to allow the worker to perform the job safely
- results of medical exams
- information sent to the physician from the employer

OSHA requires a long retention time for these documents to prove no effect from exposure over extended periods of time. Some chronic exposures do not show results until well after the exposure.

Glossary

Accident: An unexpected event generally resulting in injury, loss of property, or disruption of service.

Action Level: A quantitative limit of a chemical, biological, or radiological agent at which actions are taken to prevent or reduce exposure or contact.

Acute Exposure: A dose that is delivered to a receptor in a single event or in a short period of time.

Air Surveillance: Use of air monitoring and air sampling during a response to identify and quantify airborne contaminants on- and off-site and to monitor changes in air contaminants that occur over the lifetime of the incidents.

Chronic Exposure: Low doses repeatedly delivered to a receptor over a long period of time.

Confinement: Control methods used to limit the physical area or size of a released material. Examples: dams, dikes, and absorption processes.

Containment: Control methods used to keep the material in its container. Examples: plugging and patching.

Contaminant/Contamination: An unwanted and nonbeneficial substance.

Control: Chemical or physical methods used to prevent or reduce the hazards associated with a material. Example: neutralizing an acid spill.

Decontamination: The process of physically removing contaminants from individuals and equipment or changing their chemical nature to innocuous substances.

Degree of Hazard: A relative measure of how much harm a substance can do.

Direct-Reading Instruments: A portable device that rapidly measures and displays the concentration of a contaminant in the environment.

Emergency: A sudden and unexpected event calling for immediate action.

Emergency Removal: Action or actions undertaken, in a time-critical situation, to prevent, minimize, or mitigate a release that poses an immediate and/or significant threat(s) to human health or welfare or to the environment. (See also **Removal Action.**)

Environmental Assessment: The measurement or prediction of the concentration, transport, dispersion, and final fate of a released hazardous substance in the environment.

Environmental Emergencies: Incidents involving the release (or potential release) of hazardous materials into the environment which require immediate action.

Environmental Hazard: A condition capable of posing an unreasonable risk to air, water, or soil quality, and to plants or wildlife.

First Responder: The first personnel to arrive on the scene of a hazardous materials incident. Usually officials from local emergency services, firefighters, and police.

Hazard: A circumstance or condition that can do harm. Hazards are categorized into four groups: biological, chemical, radiation, and physical.

Hazard Classes: A series of nine descriptive terms that have been established by the UN Committee of Experts to categorize the hazardous nature of chemical, physical, and biological materials. These categories are the following:

1. Explosive
2. Nonflammable and flammable gases
3. Flammable liquids
4. Flammable solids
5. Oxidizing materials
6. Poisons, irritants, and disease-causing materials
7. Radioactive materials
8. Corrosive materials
9. Dangerous materials

Hazard Evaluation: The impact or risk the hazardous substance poses to public health and the environment.

Hazardous: Capable of posing an unreasonable risk to health and safety (Department of Transportation). Capable of doing harm.

Hazardous Material: A substance or material which has been determined by the Secretary of Transportation to be capable of posing an unreasonable risk to health, safety, and property when transported in commerce, and which has been so designated. (Department of Transportation)

Hazardous Sample: Samples that are considered to contain high concentrations of contaminants.

Hazardous Substance: 1. A material and its mixtures or solutions that are listed in the Appendix to the Hazardous Materials Table, 49 CFR 172.101, when offered for transportation in one package, or in one transport vehicle if not packaged, and when the quantity of the material therein equals or exceeds the reportable quantity. **2.** Any substance designated pursuant to Section 311(b)(2) (A) of the Federal Water Pollution Control Act; **3.** any element, compound, mixture solution, or substance designated pursuant to Section 102 of the Act; **4.** any hazardous water having the characteristics identified under or listed pursuant to Section 3001 of the Solid Waste Disposal Act (but not including any waste, the regulation of which under the Solid Waste Disposal Act has been suspended by Act of Congress); **5.** any toxic pollutant listed under Section 307(a) of the Federal Waste Pollution Control Act; **6.** any hazardous air pollutant listed under Section 112 of the Clean Air Act; and **7.** any imminently hazardous chemical substance or mixture with respect to which the Administrator has taken action pursuant to Section 7 of the Toxic Substances Control Act. The term does not include petroleum, including crude oil or any fraction thereof which is not otherwise specifically listed or designated as a hazardous substance under subparagraphs **1** through **7** of this paragraph, and the term does not include natural gas, natural gas liquids, liquefied natural gas, or synthetic gas usable for fuel (or mixtures of natural gas and such synthetic gas).

Hazardous Waste: Any material that is subject to the hazardous waste manifest requirements of the Environmental Protection Agency specified in 40 CFR, Part 262, or that would be subject to these requirements in the absence of an interim authorization to a State under 40 CFR Part 123, Subpart F.

Incident: The release or potential release of a hazardous substance or material into the environment.

Incident Characterization: The process of identifying the substance(s) involved in an incident, determining exposure pathways and projecting the effect it will have on people, property, wildlife and plants, and services.

Incident Evaluation: The process of assessing the impact that released or potentially released substances pose to public health and the environment.

Information: Knowledge acquired concerning the conditions or circumstances particular to an incident.

Inspection: Same as **Investigation.**

Intelligence: Information obtained from existing records or documentation, placards, labels, signs, special configuration of containers, visual observations, technical records, eye witnesses, and others.

Investigation: On-site and off-site survey(s) conducted to provide a qualitative and quantitative assessment of hazards associated with a site.

Limited Quantity: With the exception of Poison B materials, the maximum amount of a hazardous material for which there is a specific labeling and packaging exception.

Mitigation: Actions taken to prevent or reduce the severity of threats to human health and the environment.

Monitoring: The process of sampling and measuring certain environmental parameters on a real-time basis for spatial and time variations. For example, air monitoring may be conducted with direct-reading instruments to indicate relative changes in air contaminant concentrations at various times.

National Contingency Plan: Policies and procedures that the Federal Government follows in implementing responses to hazardous substances.

Off-Site: Outside the boundaries of the worksite.

On-Site: Within the boundaries of the worksite.

Pathways of Dispersion: The environmental medium (water, groundwater, soil, and air) through which a chemical is transported.

Persistent Chemicals: A substance which resists biodegradation and/or chemical transformation when released into the environment and which tends to accumulate on land, in air, in water, or in organic matter.

Pollutant: A substance or mixture which, after release into the environment, may cause adverse effects in organisms or their offspring.

Pollutant Transport: An array of mechanisms by which a substance may migrate outside the immediate location of the release or discharge of the substance. For example, pollution of groundwater by hazardous waste leachate migrating from a landfill.

Qualified Individual: A person who through education, experience, or professional accreditation is competent to make judgments concerning a particular subject matter. A Certified Industrial Hygienist may be a qualified individual for preparing a site safety plan.

Regulated Material: A substance or material that is subject to regulations set forth by the Environmental Protection Agency, the Department of Transportation, or any other federal agency.

Release: Any spilling, leaking, pumping, pouring, emitting, emptying, discharging, injecting, escaping, leaching, dumping, or disposing of hazardous substances into the environment.

Reportable Quantity: As set forth in the Clean Water Act, the minimum amount (pounds or kilograms) of a hazardous substance that may be discharged in a 24-hour period that requires notification of the appropriate government agency.

Response Actions: Actions taken to recognize, evaluate, and control an incident.

Response Operations: Same as **Response Actions**.

Risk: The probability that harm will occur.

Risk Assessment: The use of factual base to define the health effects of exposure of individuals or populations to hazardous materials and situations.

Risk Management: The process of weighing policy alternatives and selecting the most appropriate regulatory action integrating the results of risk assessment with engineering data and with social and economic concerns to reach a decision.

Routes of Exposure: The manner in which a contaminant enters the body through inhalation, ingestion, skin absorption, and injection.

Safety: Freedom from human, equipment, material, and environmental interactions that result in injury or illness.

Sampling: The collection of a representative portion of the universe. Example: the collection of a water sample from a contaminated stream.

Severe: A relative term used to describe the degree to which hazardous material releases can cause adverse effects to human health and the environment.

Site: Location.

Toxicity: The ability of a substance to produce injury once it reaches a susceptible site in or on the body.

Acronyms

ACGIH: American Conference of Governmental Industrial Hygienists

AIHA: American Industrial Hygiene Association

ANSI: American National Standards Institute

APF: Assigned protection factor

APR: Air-purifying respirator

ASR: Atmosphere-supplying respirator

ASTM: American Society for Testing and Materials

BEIs: Biological exposure indices

BOD: Biological oxygen demand

B of M: Bureau of Mines

Btu: British Thermal Unit

C: Ceiling

CAG: Carcinogen Assessment Group

CDC: Centers for Disease Control

CERCLA: Comprehensive Environmental Response, Compensation and Liability Act (1980)

CFR: Code of Federal Regulations

CGI: Combustible gas indicator

CHEMTREC: Chemical Transportation Emergency Center

CHRIS: Chemical Hazard Response Information System

CMA: Chemical Manufacturers' Association

CPC: Chemical protective clothing

CPE: Chlorinated polyethylene

CPM: Counts per minute

CRP: Community relations plan

DDT: Dichlorodiphenyltrichloroethane

DECON: Decontamination

DFM: Diesel fuel marine

DHHS: U.S. Department of Health and Human Services

DOD: U.S. Department of Defense

DOI: U.S. Department of the Interior

DOL: U.S. Department of Labor

DOT: U.S. Department of Transportation

DRI: Direct-reading instruments

EERU: Environmental Emergency Response Unit

EL: Exposure limit

EPA: U.S. Environmental Protection Agency

ERCS: Emergency Response Cleanup Services (under EPA contract)

ERT: Environmental Response Team

eV: Electron volt

FEMA: Federal Emergency Management Agency

FES: Fully encapsulating suit

FID: Flame ionization detector

FIT: Field Investigation Team (under contract to EPA)

HASP: Health and safety plan

HazCom: Federal Hazard Communications Standard

HEPA: High-efficiency particulate air filter (common use: "HEPA filter")

HMIS: Hazardous Materials Identification System

IDLH: mmediately dangerous to life or health

IUPAC: International Union of Pure and Applied Chemists

LC_{50}: Lethal concentration, 50 percent

LD_{50}: Lethal dose, 50 percent

LC_{Lo}: Lethal concentration—low

LD_{Lo}: Lethal dose—low

LEL: Lower explosive limit

LFL: Lower flammable limit

MACs: Maximum allowable concentrations

mg/L: Milligrams per liter

mg/m^3: Milligrams per cubic meter

MSDS: Material safety data sheets

MSHA: Mine Safety and Health Administration

NCP: National Contingency Plan

NEC: National Electrical Code

NFPA: National Fire Protection Association

NIOSH: National Institute for Occupational Safety and Health

NOAA: National Oceanic and Atmospheric Administration

NOS or n.o.s.: Not otherwise specified

NPL: National Priorities List

NRC: Nuclear Regulatory Commission

NRT: National Response Team

OHMTADS: Oil and Hazardous Materials Technical Assistance Data System

ORM: Other regulated material (specific classes such as ORM-A, ORM-E, etc.)

OSC: On-scene coordinator

OSHA: Occupational Safety and Health Administration

OVA: Organic vapor analyzer

OVM: Organic vapor meter

PCB: Polychlorinated biphenyl

PEL: Permissible exposure limit

PF: Protection factor

PID: Photoionization detector

Ppb: Parts per billion

PPE: Personal protective equipment

Ppm: Parts per million

Ppt: Parts per trillion

QA/QC: Quality assurance and quality control

RCRA: Resource Conservation and Recovery Act

REL: Recommended exposure limits

RI/FS: Remedial investigation and feasibility study

RPF: Required protection factor

RRP: Regional response plan

RRT: Regional Response Team

SAR: Supplied-air respirator

SCBA: Self-contained breathing apparatus

SOPs: Standard operating procedures

SpG: Specific gravity

STEL: Short-term exposure limit

TAT: Technical Assistance Team (under contract to EPA)

TC$_{Lo}$: Toxic concentration—low

TD$_{Lo}$: Toxic dose—low

THR: Toxic hazard rating

TLVs: Threshold limit values

TWA: Time-weighted average

UEL: Upper explosive limit

UFL: Upper flammable limit

UL: Underwriters Laboratories

UN: United Nations

USCG: United States Coast Guard

USGS: United States Geological Survey

Appendix A

EMERGENCY RESPONSE PLANNING RESOURCES

FEDERAL AGENCY RESOURCES

National Response System

The National Response System was created under the authority of the Comprehensive Environmental Response, Compensation, and Liability Act of 1980 (CERCLA) which required the development of the National Oil and Hazardous Substances Pollution Contingency Plan (commonly known as the National Contingency Plan or NCP). The purpose of the plan is to provide the federal organizational structure and procedures for preparing for and responding to discharges of oil and releases of hazardous substances. The plan establishes three organizational levels: the National Response Team (NRT), Regional Response Teams (RRTs), and OnScene Coordinators (OSCs).

NRT: A national planning, policy, and coordinating body consisting of 14 federal agencies with interests and expertise in emergency response to oil discharges and hazardous substance releases.

RRTs: Regional planning, policy, and coordinating bodies located in the ten federal regions, the Caribbean, Pacific Oceania, and Alaska. RRT membership parallels NRT membership with the addition of a representative from each state in the region. Neither the NRT nor the RRTs respond directly to incidents although they provide technical advice to an OSC and have access to resources (e.g., equipment) during an incident. Three Joint Response Teams have also been established to promote international planning and coordination with Canada, Mexico, and the Russian Federation.

OSC: A federal official predesignated by the Environmental Protection Agency for inland areas and the U.S. Coast Guard for coastal areas. The OSC coordinates all federal containment, removal, and disposal efforts and resources during an incident. The Department of Energy also has designated OSCs for dealing with any releases from their facilities.

Through the National Response Team and Regional Response Teams (RRT), federal agencies are working to combine their exercise resources, share information, broaden exercises to include hazardous materials scenarios, and expand exercise involvement to include all interdisciplinary elements. Each RRT is co-chaired by the Environmental Protection Agency (EPA) and the United States Coast Guard (USCG). To contact the RRT Co-Chairs, utilize the FEMA-DOT Hazardous Materials Information Exchange (HMIX) for an up-to-date list of names, addresses, and telephone numbers.

The NRT, RRTs, and OSCs, together with the National Response Center (NRC), form the National Response System, which is responsible for the overall coordination of federal activities related to oil discharges and hazardous materials releases. The National Response Center, a central point for receiving incident notifications and collecting incident information, provides technical data to support OSCs in response to an incident.

The National Contingency Plan also establishes requirements for Federal regional and OSC contingency plans. A regional contingency plan must be developed by each RRT as a means for coordinating timely, effective responses by federal agencies and other organizations to oil discharges and hazardous substance releases. An OSC contingency plan may be developed for responses in each OSC's area of responsibility. OSC contingency plans should be compatible with all appropriate response plans of state, local, and other non-federal entities.

The National Response Center toll-free telephone number for reporting oil and hazardous substance releases is **1-800-424-8802**.

Federal Emergency Management Agency

The Federal Emergency Management Agency (FEMA) provides a number of resources to assist state and local officials in designing, conducting, and evaluating emergency exercises.

Regional Offices

FEMA Regional Office Hazardous Materials Program Staff members are available, as time permits, to assist state and local governments in all aspects of planning, conducting, and evaluating exercises. Moreover, an Exercise Specialist in each FEMA Regional Office serves as a focal point for scenario development, pre-exercise training, and post-exercise evaluation.

FEMA also supports a State Training Officer and an Exercise Training Officer (ETO) in almost all State Emergency Management Offices. The ETOs are available to aid local communities by furnishing materials, planning exercises, conducting pre-exercise training, evaluating exercises and preparing after-action reports. Exercise Training Officers meet once a year at FEMA's National Emergency Training Center in Emmitsburg, Maryland, to discuss exercise issues. The Exercise Training Officers are usually responsible for coordinating the State and Local Exercise Annex (SLE) required under FEMA's CCA.

For information on these or other resources FEMA has to offer, contact your FEMA Regional Office.

- Federal Emergency Management Agency
  Hazardous Materials Branch
  State and Local Programs and Support Directorate
  500 C Street, S.W.
  Washington, D.C. 20472
  202/646-2860
  FTS/876-2860

- Federal Emergency Management Agency
  National Emergency Training Center
  16825 South Seton Avenue
  Emmitsburg, MD 21727
  301/447-1000
  FTS/652-1000

- HAZARDOUS MATERIALS INFORMATION EXCHANGE (HMIX)
  Electronic Bulletin Board: 708/972-3275 or FTS/972-3275
  Toll-Free Access Number: 1-800-874-2884
  Toll-free Assistance Number: 1-800-PLANFOR(1-800-752-6367)
  (In Illinois: 1-800-367-9592)

REGION I FEMA
Room 462
J.W. McCormack Post Office
& Courthouse Building
Boston, MA 02109-4595
617/223-4412
FTS/223-4412

REGION II FEMA
Room 1351
26 Federal Plaza
New York, NY   10278
212/238-8225
FtS/649-8225

REGION III FEMA
Second Floor
Liberty Square Building
105 South Seventh St.
Philadelphia, PA 19106
215/931-5528
FTS/489-5528

REGION VI FEMA
Federal Regional Center
800 North Loop 288
Denton, TX 76201-3608
817/898-9137
FTS/749-9137

REGION VII FEMA
Room 200
911 Walnut St.
Kansas City, MO 64106
816/283-7011
FTS/759-7011

REGION VIII FEMA
Denver Federal Center
Building 710, Box 25267
Denver, CO 80225-0267
303/235-4923
FTS/322-4923

REGION IV FEMA
Suite 700
1371 Peachtree St. N.E.
Atlanta, GA 30309
404/853-4454
FTS/230-4454

REGION V FEMA
Fourth Floor
West Jackson Blvd.
Chicago, IL 60604-2698
312/408-5524
FTS/363-5524

REGION IX FEMA
Building 105
Presidio of San Francisco, CA 94129
415/923-7187
FTS/469-7187

REGION X FEMA
Federal Regional Center
130 228th St., S.W.
Bothell, WA 98021-9796
206/487-4696
FTS/390-4696

Environmental Protection Agency

The Environmental Protection Agency (EPA) offers a range of resources and assistance for hazardous materials exercises. EPA Regional Offices play an integral part in working with state and local officials to ensure effective exercises are conducted. If a state or local community intends to exercise a SARA Title III plan, the Regional Office Chemical Emergency Preparedness and Prevention (CEPP) Coordinators are available to provide assistance and advice. Two sources of direct technical assistance in conducting exercises include the Environmental Response Team (ERT) and contractor support, particularly from the EPA Technical Assistance Teams (TATs).

EPA's Environmental Response Team (ERT), located in Cincinnati, OH, and Edison, NJ, is a group of highly trained scientists and engineers having expertise in multimedia sampling and analysis, hazard evaluation, environmental assessment, and cleanup techniques. The ERT offers assistance in conducting full-field exercises, coordinating the effort with EPA Regional Offices. The ERT works with a community to design a scenario relevant to the local situation and serves as a facilitator in carrying out the exercise. A debriefing is held on the day following an exercise which allows participants to evaluate their roles and to identify areas and gaps in planning activities and response capabilities which need to be addressed. The ERT, which is available to provide overall technical support to On-Scene Coordinators (OSCs) in actual incidents, conducts approximately ten full-field exercises a year.

■ Environmental Protection Agency
Chemical Emergency Preparedness and prevention Office
Office of solid Waste and Emergency Response
401 M Street, S.W.
Washington, D.C. 20460
202/475-8600

■ Environmental Protection Agency
Emergency Response Team
26 West St. Clair Street
Cincinnati, OH 45268
513/569-7537
FTS/684-7537

■ Environmental Protection Agency
Emergency Response Team
Woodbridge Ave.
Edison, NJ 08837
201/321-6740
FTS/340-6740

■ Environmental Protection Agency
Emergency Planning and Community Right-to-Know
Information Hotline
1-800-535-0202
1-202-479-2449 in Washington, D.C.

Environmental Protection Agency—Agency Regional Offices

REGION I EPA
New England Regional Laboratory
Laboratory
60 Westview St.
Lexington, MA 02173
617/860-4300
FTS/860-4300, ext. 221

REGION II EPA
Woodbridge Ave.
Edison, NJ 08837
201/321-6656
FTS/321-6656

REGION III EPA
841 Chestnut St.
Philadelphia, PA 19107
215/597-0922
FTS/597-0922

REGION VI EPA
Allied Bank Tower
1445 Ross Ave
Dallas, TX 75202-2733
214/655-2270
FTS/255-2270 or 2277

REGION VII EPA
726 Minnesota Avenue
Kansas City, KS 66101
913/236-2806
FTS/757-2806

REGION VIII EPA
One Denver Place
Suite 1300 999-18th St.
Denver, CO 80202-2413
303/293-1723
FTS/564-1723

REGION IV EPA
345 Courtland St., N.E.
Atlanta, GA 30365
FTS/257-3931

REGION IX EPA, H-12
75 Hawthorne Street.
San Francisco, CA 94105
415/744-2100
FTS/484-2100

REGION V EPA
230 S. Dearborn St.
Chicago, IL 60604
312/886-1964
FTS/886-1964

REGION X EPA
1200 6th Avenue
Seattle, WA 98101
206/442-1263
FTS/399-4349

U.S. Coast Guard

The U.S. Coast Guard (USCG) sponsors, with RRT support, six On-Scene Coordinator/Regional Response Team (OSC/RRT) exercise simulation training sessions across the country on an annual basis. Five of these exercises involve coastal areas; the sixth focuses on an inland incident.

OCS/RRT exercises are comprehensive and realistic simulations of hazardous materials or oil incidents used to evaluate plans, policies, procedures, and personnel. The focus of the exercise is on the OSC/RRT relationship and the management of a major incident. Over 150 federal, state, local, and industry officials generally participate in a typical exercise. Participation in the exercise provides the opportunity for Federal predesignated OSCs and RRT members to assemble in a central location with the local response community. Industry response representatives and clean-up contractors are also involved. All actions are simulated; no equipment or personnel are dispatched.

The goal of the exercise is to allow all elements of the response community to work together. The scenario is designed to reflect actual patterns in the host community. The Coast Guard Marine Safety School at Yorktown, VA, designs the scenarios in coordination with a selected team of local agency and industry representatives. Generally, each simulation involves a six-week process from initial planning through scenario development and exercise conclusion. Exercise length is two days—eight hours of simulation activity conducted in real time and three hours of open forum debriefing. The debriefing is a means to discuss any deficiencies and necessary corrective actions as well as to reinforce positive results. Communities interested in participating in a simulation should contact the RRT or the local predesignated Federal OSC.

- U.S. Coast Guard
  Marine Environmental Response Office (T-MER)
  Marine Safety School
  Reserve Training Center
  Yorktown, VA 23690
  FTS/827-2335

Department of Transportation

The Department of Transportation/Research and Special Programs Administration (DOT/RSPA) has a variety of program resources and technical assistance which can support the development of comprehensive hazardous materials exercises, with particular emphasis on transportation issues.

RSPA's primary source of hazardous materials transportation data, the Hazardous Materials Information System (HMIS), can be used to identify either individual incident reports or compilations of state incident history. The actual performance data derived from reports by carriers whenever there is an unintentional release of hazardous materials can be a useful source of scenario material. Individual incident information includes types of vehicles and materials involved, deaths and injuries, if any, resulting from the incident, losses and property damage, estimated cost of decontamination, and nature of packaging failure. New emphasis has been placed on enhancing HMIS use; RSPA encourages requests for data.

Information emanating from enforcement activity can be used both in determining objectives for exercises and in setting up scenarios based on identified patterns of chemicals being shipped on major transportation routes.

■ Department of Transportation
  Research and Special Programs Administration
  Federal, State, and Private Sector Initiatives Div.
  400 7th Street, S.W.
  Washington, DC 20590
  202/366-4900

Department of Health and Human Services

Under the provisions of CERCLA, the Agency for Toxic Substances and Disease Registry (ASTDR) in the Department of Health and Human Services (DHHS) is responsible for providing support to state and local governments in health matters relating to releases or potential releases of hazardous materials. ATSDR can furnish technical support to federal, state, and local agencies in planning hazardous materials exercises, and in testing and evaluating the health components of their emergency plans. ATSDR has participated in developing, staging, and evaluating both tabletop and full-scale exercises. Depending on the extent of the exercise, ATSDR input will address contamination reduction and decontamination activities related to response personnel, emergency medical services, and hospital emergency rooms. Exercises requiring decision making related to overall public health are also encouraged.

■ Department of Health and Human Resources
  U.S. Public Health Service
  Agency for Toxic Substances and Disease Registry
  600 Clifton Road (Mailstop E-28)
  Atlanta, GA 30333
  24-Hour Number: 404/639-0615
  FTS/236-0615

Department of Commerce

The National Oceanic and Atmospheric Administration (NOAA) of the Department of Commerce (DOC) currently makes available operational models for spill responses (spill trajectory and air dispersion) which can be used in exercises. For example, the Computer-Aided Management of Emergency Operations (CAMEO) system, developed by NOAA in cooperation with the fire department of Seattle, WA, is receiving increasing use by local emergency management organizations. The CAMEO system is a program jointly managed by NOAA and EPA. CAMEO is distributed by the Environmental Health Center of the National Safety Council in Washington, D.C.

CAMEO is designed to help emergency planners and first responders both plan for, and safely handle, chemical accidents. CAMEO contains response information and recommendations for 2,629 commonly transported chemicals, an air-dispersion model to assist in evaluating release scenarios and evaluation options, and several easily adaptable databases. It also provides a computational program that addresses the emergency planning provisions of Title III, the Emergency Planning and Community Right-to-Know Act of 1986. CAMEO can be used for tabletop exercises and simulations and for hazards analysis training.

CAMEO can include such diverse information as facility floor plans with chemical storage locations; contacts lists; locations of schools, hospitals, and other population concentrations; response resources; and maps of the planning area overlaid with plumes calculated by the air model.

CAMEO operates with both Macintosh and MS-DOS (IBM-compatible) computers; both computers programs are functionally equivalent.

■ Department of Commerce
   National Oceanic and Atmospheric Administration
   Hazardous Materials Response Branch
   7600 Sand Point Way, N.E.
   Seattle, WA 98115
   206/536-6317
   FTS/392-6317

Department of Defense

Each military service in the Department of Defense (DOD) has well-established programs for routine testing of emergency response plans.

The U.S. Army Material Command Surety Field Activity, for example, is responsible for providing technical supervision of the Army Material Command's chemical surety, nuclear, and nuclear reactor facility accident/incident response and assistance activities, including exercising of the Army Service Response Force. Army installations/organizations that have the mission of storing, handling, or using military hazardous materials are required to develop contingency or operation response and assistance plans. Additionally, they are required to conduct quarterly exercises. One of these exercises should involve testing existing state, local, or other supporting agency plans on an annual basis.

Every two years, the Army through the Army Material Command conducts an Army-wide exercise to test and improve the Army's capability to respond to an incident involving chemical surety materials. The first of these exercises included FEMA and EPA as participants. As a result of this exercise, it was recommended that offsite response considerations be extended to include greater participation from state and local government representatives and to focus greater attention on the needs of evacuated civilians. The exercise also emphasized the need for clarification of OSC designation during an incident at a defense facility.

■ Department of the Army
U.S. Army Material Command
Surety Field Activity
Picatinny Arsenal, NJ 07806-5000
201/724-4836

PRIVATE SECTOR RESOURCES

In addition to state and federal agency support, the private sector can provide numerous resources (e.g., technical assistance, planning capabilities, and equipment). Industry resources, when combined with local, state, and perhaps federal resources and assistance, can improve overall emergency preparedness, promote public safety, and provide for a multi-disciplinary approach to a comprehensive exercise program.

■ Association of American Railroads
Bureau of Explosives
50 F St. N.W.
Washington, D.C. 20001
202/639-2133

■ Chemical Manufacturers Association
2501 M St. N.W.
Washington, D.C. 20037
202/887-1100

■ Joint Commission on Accreditation of Healthcare Organizations
875 North Michigan Avenue
Chicago, Il 60611
312/642-6061

To contact Federal Agency and Private Sector sources of exercise assistance, utilize the FEMA-DOT Hazardous Materials Information Exchange (HMIX) at (708) 972-3275 or 1-800-874-2884. The HMIX computer bulletin board contains an up-to-date list of names, addresses, and telephone numbers as well as other resources to tap prior to an exercise. The HMIX is one of the best ways to stay abreast of available exercise resources.

Appendix B

EMPLOYEE TRAINING RECORD (SAMPLE)

EMPLOYEE TRAINING RECORD

Employee Name	Employee No.	HAZCOM	Lockout/ Tagout	HAZWOPER	PSM	First Aid	Next HAZCOM	Next Lockout/ Tagout	Next HAZWOPER	Next PSM	Next First Aid
B. E. Corl	1224	12/5/96	10/20/96	11/2/96	1/17/97	11/2/96	12/5/97	10/20/97	11/2/97	1/17/98	11/2/97
C. LaPenta	1405	12/5/96	10/20/96	11/2/96	1/17/97	11/2/96	12/5/97	10/20/97	11/2/97	1/17/98	11/2/97
J. Kennedy	1178	12/5/96	10/20/96	11/2/96	1/17/97	11/2/96	12/5/97	10/20/97	11/2/97	1/17/98	11/2/97
M. Huber	1334	12/5/96	10/20/96	11/2/96	1/17/97	11/2/96	12/5/97	10/20/97	11/2/97	1/17/98	11/2/97
R. Brachfeld	1203	12/5/96	10/20/96	11/2/96	1/17/97	11/2/96	12/5/97	10/20/97	11/2/97	1/17/98	11/2/97
T. James	1377	12/5/96	10/20/96	11/2/96	1/17/97	11/2/96	12/5/97	10/20/97	11/2/97	1/17/98	11/2/97

Appendix C

EMERGENCY RESPONSE NOTIFICATION FORM

PART 1

Corporate Crisis Hotline:	1-800-
National Response Center:	1-800-424-8802
Local Emergency Coordinator (City):	
Local Emergency Coordinator (County):	
State Emergency Coordinator:	
Facility Coordinator:	
Fire Department:	911 or local number
Police Department:	911 or local number

PART 2

The following script should be followed when making agency notifications.

This is the *FACILTY*, at *(state your location)*.

My name is *(state your name)*.

I am the *(insert your position at facility)*, and my telephone number is *(insert facility phone number and your extension number, if any)*.

I am calling to report a release of *(insert name of material)*.

This leak occurred at *(insert time and date)* and has/has not been contained as of this moment.

OR

This leak occurred at *(insert time and date)* and is ongoing and is not expected to be contained/stopped until *(estimate time leak will be stopped)*.

This is a *(choose one)*:

- **Site Emergency**:   Release has occurred and will probably not have an off-site impact.
- **General Emergency**: Release has occurred which will probably have an off-site impact.

The estimated quantity of *(insert name of material)* released is *(insert quantity or unknown)*.

The current weather conditions, as measured at the plant, are a wind speed of *(insert speed)* in a direction that is *(insert from-to direction)*.

We have *(insert number)* of injured personnel who *will/will not* require medical assistance.

We *need/do not need* your assistance at this time to *(describe what you need)*.

Please tell me my case number: _____
(Write number here)

Do you have any questions?

PART 3: ADDED INFORMATION FOR MANAGEMENT REPORT

Name of person making notification: _____

Names of persons notified: _____

National Response Center: Date: Time:

   Individual: Case #:

Local Emergency Coordinator: Date: Time:

   Individual:

State Emergency Response: Date: Time:

   Individual:

Fire Department: Date: Time:

   Individual:

Police Department: Date: Time:

   Individual:

Other (   ):   Date: Time:

   Individual:

Other (   ):   Date: Time:

   Individual:

Appendix D

SAMPLE SPILL PREVENTION, CONTROL, AND COUNTERMEASURE PLAN

METROPOLITAN WATER RECLAMATION DISTRICT OF GREATER CHICAGO

Elements of a Spill Prevention, Control and Countermeasure Plan

I. General Information

A. Company Name, Address and Telephone Number

B. Emergency Contact Personnel

    1. Primary Emergency Coordinator
        a. Name
        b. Title
        c. Work Phone Number
        d. 24-Hour Emergency Phone Number

C. Description of Business

    1. Standard Industrial Classification Number
    2. Applicable Categorical Pretreatment Regulations
    3. Days/Shifts/Hours of Operation
    4. Number of Employees
    5. Description of Operations
    6. Description of Pretreatment Practices

D. Discharge Practices

    1. Average Daily Discharge (gpd)
    2. Chemical Constituents of Discharge
    3. Nature of Discharge

        a. Continuous
            (1) Daily Discharge Volume
            (2) Constituents of Discharge

        b. Batch
            (1) Frequency of Batch Discharge
            (2) Volume of Batch Discharge
            (3) Constituents of Batch Discharge

II. Plant Layout/Flow Diagrams

A. General Plant Layout

    1. Property Boundaries
    2. Entrances/Exits
    3. Manufacturing Area

4. Loading/Unloading Area
5. Hazardous Material Storage Area
6. Pretreatment Facilities
7. Waste Handling/Storage Area
8. Security or Warning System Signs

B. Flow Drainage Diagram

1. Identification and location of all floor drains, drainage pipes and channels and their direction of flow (see Element IV, B, 1 for drainage prohibitions)
2. Identification and location of final wastewater discharge points, sumps or storage tanks
3. Direction of flow for all sanitary/storm sewers
4. Direction of above ground run-off from:

   (a) Chemical Storage Area
   (b) Pretreatment Facilities
   (c) Waste Handling Area

III. Material Inventory

A. Description and location of stored chemicals, production residues and sludges

B. Maximum quantity of stored chemicals, production residues and sludge on hand

C. Description of storage containers, container attachments and transfer equipment

D. Chemical compatibility of stored material with storage containers and other materials stored in the immediate vicinity

IV. Spill and Leak Prevention

A. Equipment to Prevent or Detect Spills (*e.g.*, holding tanks, pumping equipment, underground seepage protection, cathodic protection of underground tanks, liquid level sensing devices, drip pans, overflow alarms, pH excursion alarms, ORP alarms, collision protection structures, explosion and fire prevention provisions, etc.)

B. Drainage and Secondary Containment (Absolute Requirements)

1. <u>No floor drains or other direct bypass to the sewer system may exist in wet manufacturing areas, wastewater pretreatment areas, or raw chemical/sludge storage areas which in the event of run-off spillage would result in a violation of any criteria of the Sewage and Waste Control Ordinance</u>

2. Under those situations when floor drains are required by specific building codes or when the elimination of floor drains is not feasible, a detailed explanation of the prohibitive factors along with an acceptable alternative plan to prevent run-off spillage from entering the sewer system must be provided

3. Adequate secondary containment, such as but not limited to impervious diking, must be provided for all manufacturing, pretreatment operations and raw chemical/sludge storage areas defined by Item No. IV, B, 1

4. Impervious flooring with no direct drainage to the sewer system must be provided for all secondary containment areas

5. A minimum secondary containment capacity of 100% of the volume of the largest above ground process tank located in wet manufacturing areas which, in the event of run-off spillage, would result in a violation of any criteria of the Sewage and Waste Control Ordinance must be provided

6. A minimum secondary containment capacity of 100% of the volume of the largest above ground wastewater pretreatment system process tank which, in the event of run-off spillage, would result in a violation of any criteria of the Sewage and Waste Control Ordinance must be provided

7. A minimum secondary containment capacity of 100% of the maximum volume of each of the following: (a) stored chemicals, (b) production residues, (c) pretreatment sludge must be provided. Outside storage facilities must be covered to prevent storm water from entering secondary containment areas

8. Supporting documentation pursuant to Item Nos. Iv, B, 1 through 7 must be submitted. Said documentation must include written narratives, sketches, and engineering computations on tank sizing, worst-case scenario spill volumes and secondary containment capacities. Further, said documentation must be prepared by a professional engineer registered in the state of Illinois and must certify that the design and capacities of the secondary containment facilities will satisfy District requirements and provide adequate protection from run-off spillage entering the sewer system

C. Preventive Maintenance Procedures and Schedules

D. Inventory of Pretreatment System Spare Parts (*e.g.*, valves, pumps, gaskets, valve packing material, sensor probes, filters, etc.)

V. Emergency Response Equipment and Procedures

A. Emergency Response Equipment (*e.g.*, alarm and communication systems, sewer plugs, sorbent materials, fire extinguishers, ventilation and breathing equipment, protective clothing, first aid kits, etc.)

B. Emergency Response Procedures

1. Notification to in-house emergency response coordinator on a 24-hour basis

2. Designated chain of command listing names, titles and telephone numbers for contact on a 24-hour basis

3. Posted listing of spill/emergency response agencies with telephone numbers and applicable notification procedures

4. Procedures for stopping flow (*e.g.*, shutting off water supply, shutting off pumps, closing influent/effluent valves, plugging outlets, etc.)

5. Description of site remediation procedures
    (a) If cleanup is performed with in-house resources, describe treatment/disposal methodologies
    (b) If cleanup is contracted to an outside party, list names and phone numbers of all contractors and consultants as well as names and night and daytime telephone numbers of company personnel with the authority to commit the company to financial participation in cleanup and remediation

VI. Spill Reporting Procedures

A. Describe procedures for reporting and documenting spills and slug discharges

1. Reporting procedures must conform with Article V, Section 4 of the Sewage and Waste Control Ordinance of the Metropolitan Water Reclamation District of Greater Chicago

B. Provide listings of all agencies to be notified in the event of a spill of slug discharge, include telephone number

VII. Employee Training Program

A. Outline of Employee Training Program
    1. Plant Processes
    2. Toxic/hazardous Material Usage
    3. Potential Safety Hazards
    4. Practices for Preventing Spills/slug Discharges
    5. Procedures for Responding to Spills/slug Discharges

B. Schedule for Employee Training

C. Maintain Records on All Employee Training

VIII. SPCC Plan Certification

A. Each SPCC Plan must be certified by a professional engineer registered in the state of Illinois and indicate that the plan will provide adequate protection from spills and slug loadings when used and maintained properly, and that the plan and containment facilities conform to all applicable federal, state, county and municipal regulations

B. Each SPCC Plan must be certified by an officer of the company and indicate that the plan has been implemented

Appendix E

RMP AND PSM APPLICABILITY QUESTIONNAIRE

GENERAL INFORMATION

Facility Name _____

Location/Address _____

Facility Contact _____ Phone Number _____

Site Description

Size of facility (sq. ft.) _____ Size of facility site (acres) _____

Is the facility in a rural area, urban area, suburban area, other? _____

Surrounding Land Use

North _____

South _____

East _____

West _____

Name and distance to nearest surface water _____

Name and distance to nearest public meeting place such as a school, church, jail or hospital

_____

Are there public lands near the facility: YES / NO

If YES, check and describe below.

_____ Recreational _____

_____ Hunting _____

_____ Fishing _____

_____ Swimming _____

_____ Parks _____

_____ Other _____

PROCESS INFORMATION

Employee Information

Number of employees _____

Number of shifts per day _____

Process Information

Describe the function of this facility _____

_____

_____

_____

SIC Code _____

	Yes	No	N/A
1. Does the facility have any of the regulated chemicals in amounts which exceed the RMP thresholds? *Include all types of storage (tank, tote, or drum). (See Appendix H.)*	❏	❏	❏
2. Does the facility have more than the RMP threshold quantity of chemicals in a process? *("Process," includes all piping, vacuum, atmospheric- or pressure-storage vessels, and charge to mixing, blending, or reaction tank).*	❏	❏	❏
3. Within the last five years, has the facility had an accidental release with offsite consequences?	❏	❏	❏
4. Does the facility have more than 10,000 pounds of any substance with a National Fire Protection Association flammability hazard rating of 4?	❏	❏	❏
5. Does the facility have any of the PSM-listed chemicals in amounts which exceed the thresholds? *Include all types of storage (tank, tote, or drum).*	❏	❏	❏
6. Does the facility have more than the threshold quantity of PSM chemicals in a process? *(For OSHA purposes, the "process," includes all piping, vacuum, atmospheric- or pressure-storage vessels, and charge to mixing, blending or reaction tank).*	❏	❏	❏
7. Does the facility have a written employee participation plan?	❏	❏	❏
8. Does the written employee participation plan cover new, current, and contract employees?	❏	❏	❏
9. Does the employee participation plan include a training program for all employees regarding process safety management?	❏	❏	❏
10. Has the facility prepared detailed Process Safety Information for all processes covered under the PSM standard?	❏	❏	❏

	Yes	No	N/A
11. Has the facility completed a Process Hazards Analysis (PHA) for all covered processes at the facility?	❏	❏	❏

If yes, what type of analysis was completed: (circle one)
- What-If?
- HAZOP
- Fault-tree
- FMEA

	Yes	No	N/A
12. Does the facility have written operating procedures for **all** phases of operation (start-up, normal, shutdown, emergency) and **all** maintenance for every covered process?	❏	❏	❏
13. Do all process operators receive training and subsequent certification?	❏	❏	❏
14. Do process operators receive regular refresher training and/or certification?	❏	❏	❏
15. Does the facility have a procedure to ensure that all facility contractors have an OSHA-compliant safety program?	❏	❏	❏
16. Are all contractors working in the facility trained regarding process safety, hazardous materials emergency response (HAZWOPER—First Responder Awareness level), and facility evacuation?	❏	❏	❏
17. Does the facility have a procedure for ensuring the safety of all new and modified processes prior to startup of the process?	❏	❏	❏
18. Does the facility have a written management of change process?	❏	❏	❏
19. Does the facility have a written mechanical integrity program?	❏	❏	❏
20. Does the written mechanical integrity program include standard procedures to ensure the ongoing integrity of process equipment?	❏	❏	❏

	Yes	No	N/A
21. Does the facility have a hot work permit program?	❏	❏	❏
22. Does the facility have a procedure for investigating incidents related to the covered process?	❏	❏	❏
23. Does the facility have a HAZWOPER (29 CFR 1910.120) compliant emergency response plan?	❏	❏	❏
24. Has the facility conducted internal or external compliance audits for PSM? If so, when? _____	❏	❏	❏

Appendix F

SAMPLE DECONTAMINATION PROCEDURES

OVERVIEW

The Facility Emergency Coordinator supervises the HAZMAT Team in its cleanup and decontamination activities. Although decontamination procedures vary, depending on the particular incident, this section outlines standard principles and procedures which would be common to all incidents.

APPLICABILITY

The Emergency Coordinator arranges the disposal of recovered waste and contaminated items. RCRA requires that the Emergency Coordinator provide for the treating and the storage or disposal of recovered waste, contaminated soil, surface water, or any other material that results from a release, fire, or explosion at the facility. All emergency response equipment that is listed in the emergency plan must also be cleaned and fit for its intended use before operations resume.

METHODS OF DECONTAMINATION

Methods for decontaminating buildings, structures, and equipment can be grouped according to the mode of action. The methods include physical/mechanical, chemical, and thermal means and solvent extraction. The following is a partial listing of methods:

- Asbestos abatement
- Absorption
- Bleaching

- Demolition

- Dismantling

- Dusting/vacuuming/wiping

- Encapsulation

- Grit blasting

- Hydroblasting/water washing

- Microbial degradation

- Painting/crating

- Scarification

- Solvent washing

- Steam cleaning

- Vapor-phase solvent extraction

Methods for decontaminating personal protective clothing and equipment may involve several different stages of washing or rinsing in various decontamination solutions of water. Some outer protective clothing may need to be disposed of, along with other cleanup wastes. Standard Operating Safety Guides have been published by the EPA which describe the various techniques used by emergency response personnel in decontaminating protective equipment.

PROCEDURES

Development of an overall decontamination strategy revolves around the proper identification and evaluation of the contaminants present. This knowledge is necessary for selection of decontamination methods that will effectively reduce the contamination to acceptable levels and provide adequate protection to workers and those involved in decontamination efforts.

The following procedures are conducted by the HAZMAT Team:

Conduct Visual Inspection

Before decontamination procedures begin, a determination is made of the nature and extent of the contamination by conducting a visual inspection of the area and equipment. Note: Any personnel involved in any cleanup/decontamination effort must have the proper protective equipment. The following steps are taken to develop a site-specific decontamination plan.

Determine Properties of Contaminants

Evaluate the hazards, i.e., gather information on the physical and chemical properties of the contaminants, the fire and explosion hazards, the toxicity and health hazards, and chemical reactivity (refer to MSDS). The health and safety aspects associated with the use of various cleanup techniques or processes and the exposure limits of particular contaminants are included as part of the overall hazard evaluation.

Determine Future Use

Determine the future use of contaminated buildings, structures, and equipment. This use could fall into one of three categories: unrestricted/public use, restricted or industrial use, or no future use. If unable to decontaminate to safe levels, notify the Emergency Coordinator so that he can plan an appropriate communications strategy.

Establish Target Levels

Establish target levels of decontamination when a future use has been identified.

Determine Worker Health and Safety Requirements

Determine worker health and safety requirements from the hazard evaluation data developed earlier for worker safety during decontamination.

Prepare Decontamination Plan

Prepare a site decontamination plan before cleanup is initiated. This will specify the decontamination method, the quality assurance/quality control procedures to be followed, the equipment and support facilities needed, the method of residue disposal, worker health and safety precautions, personal protective clothing, and scheduling.

Conduct Decontamination

Initiate and conduct cleanup in accordance with plan. Facility management will arrange for the disposal of cleanup debris. Once cleanup has been initiated, contaminant levels will be monitored through the course of the decontamination operation. Determine the extent of residual contamination following decontamination so that the effectiveness of the cleanup methods will be properly assessed. If contamination approaches the fence line, keep the Emergency Coordinator informed of the situation.

Decontaminate Equipment

Decontaminate equipment used in removal operations before leaving the site of contamination to minimize the spread of contamination.

Appendix G

EMERGENCY RESPONSE CRITIQUE FORM

EMERGENCY RESPONSE CRITIQUE FORM

This form is to be completed after each implementation of the Emergency Response Plan and submitted to the Corporate Safety and Health Manager within 30 days.

Date Emergency Response Plan was Implemented:	
Time Emergency Response Plan was Implemented:	

Plan was implemented for a *(please circle one)*:

        Drill   Fire   Release   Other (Specify):

Were there any problems in implementing the plan?

        Yes   No

        Comments:

Was the performance of the site personnel satisfactory?

        Yes   No

        Comments:

Was the performance of the emergency response team satisfactory?

        Yes   No

        Comments:

EMERGENCY RESPONSE CRITQUE FORM, continued

Was the treatment of exposed personnel on-site adequate?

        Yes   No   N/A

        Comments:

Was the treatment of exposed personnel off-site adequate?

        Yes   No   N/A

        Comments:

Was the on-site communications system adequate?

        Yes   No   N/A

        Comments:

Was the off-site communications system adequate?

        Yes   No   N/A

        Comments:

Were the emergency power and lighting systems adequate?

        Yes   No   N/A

        Comments:

EMERGENCY RESPONSE CRITQUE FORM, continued

Recommendations for changes in equipment, procedures, additional comments, etc.:

_____   _____
(Signature of ERC)   (Date)

Appendix H

REGULATED SUBSTANCES FOR RMP AND PSM RULES

REGULATED SUBSTANCES FOR RMP AND PSM RULES

Chemical Name	RMP Threshold Quantity (lbs)		PSM Threshold Quantity (lbs)	
Acetaldehyde	10,000	❑	2,500	❑
Acetylene	10,000	❑	N/A	
Acrolein	5,000	❑	150	❑
Acrylonitrile	20,000	❑	N/A	
Acrylyl chloride	5,000	❑	250	❑
Alkylaluminums	N/A		5,000	❑
Allyl chloride	N/A		1,000	❑
Allyl alcohol	15,000	❑	N/A	
Allylamine	10,000	❑	1,000	❑
Ammonia (aqueous sol.,conc. 20% or greater)	20,000	❑	15,000	❑
Ammonia (anhydrous)	10,000	❑	10,000	❑
Ammonium Permanganate	N/A		7,500	❑
Ammonium Perchlorate	N/A		7,500	❑
Arsenous trichloride	15,000	❑	N/A	
Arsine	1,000	❑	100	❑
Boron trifluoride compound with methyl ether(1:1)	15,000	❑	N/A	
Boron trichloride	5,000	❑	2,500	❑
Boron trifluoride	5,000	❑	250	❑
Bromine pentafluoride	N/A		2,500	❑
Bromine chloride	N/A		1,500	❑
Bromine	10,000	❑	1,500	❑
Bromine trifluoride	N/A		15,000	❑
3-Bromopropyne (also called Propargyl bromide)	N/A		100	❑
Bromotrifluorethylene	10,000	❑	10,000	❑
1,3-Butadiene	10,000	❑	10,000	❑
Butane	10,000	❑	10,000	❑
Butene	10,000	❑	10,000	❑
1 Butene	10,000	❑	10,000	❑
2 Butene cis	10,000	❑	10,000	❑
2 Butene trans	10,000	❑	10,000	❑
Butyl Hydroperoxide (Tertiary)	N/A		5,000	❑
Butyl Perbenzoate (Tertiary)	N/A		7,500	❑
Carbon disulfide	20,000	❑	10,000	❑
Carbon oxysulfide	10,000	❑	10,000	❑
Carbonyl Fluoride	N/A		2,500	❑
Cellulose Nitrate (concentration greater than 12.6% nitrogen)	N/A		2,500	❑
Chlorine Trifluoride	N/A		1,000	❑

Chemical Name	RMP Threshold Quantity (lbs)		PSM Threshold Quantity (lbs)	
Chlorine Pentrafluoride	N/A		1,000	❑
Chlorine	2,500	❑	1,500	❑
Chlorine monoxide	10,000	❑	10,000	❑
Chlorine dioxide	1,000	❑	1,000	❑
1-Chloro-2,4-dinitrobenzene	N/A		5,000	❑
Chlorodiethylaluminum (also called Diethylaluminum chloride)	N/A		5,000	❑
Chloroform	20,000	❑	N/A	
Chloromethyl methyl ether	5,000	❑	100	❑
Chloromethyl ether	1,000	❑	N/A	
Chloropicrin and Methyl bromide mixture	N/A		1,500	❑
Chloropicrin and Methyl chloride mixture	N/A		1,500	❑
Chloropicrin	N/A		500	❑
1-Chloropropylene	10,000	❑	10,000	❑
2-Chloropropylene	10,000	❑	10,000	❑
Commune Hydroperoxide	N/A		5,000	❑
Crotonaldehyde, (E)	20,000	❑	N/A	
Crotonaldehyde	20,000	❑	N/A	
Cyanogen chloride	10,000	❑	500	❑
Cyanogen	10,000	❑	2,500	❑
Cyanuric fluoride	N/A		100	❑
Cyclohexylamine	15,000	❑	N/A	
Cyclopropane	10,000	❑	10,000	❑
Diacetyl peroxide (concentration greater than 70%)	N/A		5,000	❑
Diazomethane	N/A		500	❑
Dibenzoyl peroxide	N/A		7,500	❑
Diborane	2,500	❑	100	❑
Dibutyl peroxide (tertiary)	N/A		5,000	❑
Dichloroacetylene	N/A		250	❑
Dichlorosilane	10,000	❑	2,500	❑
Diethylzinc	N/A		10,000	❑
Difluoroethane	10,000	❑	N/A	
Diisopropyl peroxydicarbonate	N/A		7,500	❑
Dilauroyl peroxide	N/A		7,500	❑
Dimethylamine	10,000	❑	N/A	
Dimethylamine, anhydrous	N/A		2,500	❑
Dimethyldichlorosilane	5,000.	❑	1,000	❑
1,1-Dimethylhydrazine	15,000	❑	1,000	❑
2,4-Dinitroaniline	N/A		5,000	❑
Ethyl nitrite	10,000	❑	5,000	❑
Ethyl methyl ketone peroxide (also Methyl ethyl ketone peroxide; conc. >60%)	N/A		5,000	❑

Chemical Name	RMP Threshold Quantity (lbs)		PSM Threshold Quantity (lbs)	
Ethyl mercaptan	10,000	❏	10,000	❏
Ethyl ether	10,000	❏	10,000	❏
Ethyl chloride	10,000	❏	10,000	❏
Ethylamine	10,000	❏	7,000	❏
Ethylene oxide	10,000	❏	5,000	❏
Ethylene	10,000	❏	10,000	❏
Ethylene fluorohydrin	N/A		100	❏
Ethylenediamine	20,000	❏	N/A	
Ethyleneimine	10,000	❏	1,000	❏
Fluorine	1,000	❏	1,000	❏
Formaldehyde	15,000	❏	1,000	❏
Furan	5,000	❏	500	❏
Hexafluoroacetone	N/A		5,000	❏
Hydrazine	15,000	❏	N/A	
Hydrochloric acid (sol.,conc 37% or greater)	15,000	❏	5,000	❏
Hydrocyanic acid	2,500	❏	1,000	❏
Hydrogen bromide	N/A		5,000	❏
Hydrogen peroxide (52% by weight or greater)	N/A		7,500	❏
Hydrogen	10,000	❏	10,000	❏
Hydrogen fluoride (conc. 50% greater)	1,000	❏	1,000	❏
Hydrogen chloride (anhydrous)	5,000	❏	5,000	❏
Hydrogen sulfide	10,000	❏	1,500	❏
Hydrogen selenide	500	❏	150	❏
Hydroxylamine	N/A		2,500	❏
Iron, pentacarbonyl	2,500	❏	250	❏
Isobutane	10,000	❏	10,000	❏
Isobutyronitrile	20,000	❏	N/A	
Isopentane	10,000	❏	10,000	❏
Isoprene	10,000	❏	10,000	❏
Isopropyl chloride	10,000	❏	10,000	❏
Isopropyl chloroformate	15,000	❏	N/A	
Isopropylamine	10,000	❏	5,000	❏
Ketene	N/A		100	❏
Methacrylaldehyde	N/A		1,000	❏
Methacrylonitrile	10,000	❏	250	❏
Methacryloyl chloride	N/A		150	❏
Methacryloyloxyethyl isocyanate	N/A		100	❏
Methane	10,000	❏	10,000	❏
2-Methyl-1-butene	10,000	❏	10,000	❏
3-Methyl-1-butene	10,000	❏	10,000	❏
Methyl chloroformate	5,000	❏	500	❏

Chemical Name	RMP Threshold Quantity (lbs)		PSM Threshold Quantity (lbs)	
Methyl ether	10,000	❏	10,000	❏
Methyl formate	10,000	❏	10,000	❏
Methyl hydrazine	15,000	❏	100	❏
Methyl isocyanate	10,000	❏	250	❏
Methyl mercaptan	10,000	❏	5,000	❏
Methyl vinyl ketone	N/A		100	❏
Methyl iodide	N/A		7,500	❏
Methyl fluorosulfate	N/A		100	❏
Methyl thiocyanate	20,000	❏	N/A	
Methyl chloride	10,000	❏	15,000	❏
Methyl fluoroacetate	N/A		100	❏
Methyl ethyl ketone peroxide (also Methyl ethyl ketone peroxide; conc. >60%)	N/A		5,000	❏
Methylamine	10,000	❏	1,000	❏
2-Methylpropene	10,000	❏	10,000	❏
Methyltrichlorosilane	5,000	❏	500	❏
Nickel carbonyl	1,000	❏	150	❏
Nitric oxide	10,000	❏	250	❏
Nitric acid	15,000	❏	500	❏
Nitroaniline (para-Nitroaniline)	N/A		5,000	❏
Nitrogen tetroxide (also called Nitrogen peroxide)	N/A		250	❏
Nitrogen trifluoride	N/A		5,000	❏
Nitrogen trioxide	N/A		250	❏
Nitrogen oxides (NO; N_2O; N_2O_4; N_2O_3)	N/A		250	❏
Nitrogen dioxide	N/A		250	❏
Nitromethane	N/A		2,500	❏
Oleum (Fuming Sulfuric Acid) [Sulfuric Acid, mixture with sulfur trioxide]	10,000	❏	1,000	❏
Osmium tetroxide	N/A		100	❏
Oxygen difluoride (Fluorine monoxide)	N/A		100	❏
Ozone	N/A		100	❏
Pentaborane	N/A		100	❏
1,3-Pentadiene	10,000	❏	10,000	❏
Pentane	10,000	❏	10,000	❏
1-Pentene	10,000	❏	10,000	❏
2-Pentene, (E)	10,000	❏	10,000	❏
2-Pentene, (Z)	10,000	❏	10,000	❏
Peracetic acid	10,000	❏	1,000	❏
Perchloric acid (conc. >60% by weight)	N/A		5,000	❏
Perchloromethylmercaptan	10,000	❏	150	❏
Perchloryl fluoride	N/A		5,000	❏
Phosgene	500	❏	100	❏

Chemical Name	RMP Threshold Quantity (lbs)		PSM Threshold Quantity (lbs)	
Phosphine	5,000	❑	100	❑
Phosphorus trichloride	15,000	❑	1,000	❑
Phosphorus oxychloride	5,000	❑	1,000	❑
Piperidine	15,000	❑	N/A	
Propadiene	10,000	❑	10,000	❑
Propane	10,000	❑	10,000	❑
Propargyl bromide	N/A		100	❑
Propionitrile	10,000	❑	N/A	
Propyl chloroformate	15,000	❑	N/A	
Propyl nitrate	N/A		2,500	❑
Propylene oxide	10,000	❑	N/A	
Propylene	10,000	❑	10,000	❑
Propyleneimine	10,000	❑	N/A	
Propyne	10,000	❑	10,000	❑
Sarin	N/A		100	❑
Selenium hexafluoride	N/A		1,000	❑
Silane	10,000	❑	10,000	❑
Stibine (Antimony Hydride)	N/A		500	❑
Sulfur tetrafluoride	2,500	❑	250	❑
Sulfur trioxide	10,000	❑	1,000	❑
Sulfur dioxide	5,000	❑	1,000	❑
Sulfur pentafluoride	N/A		250	❑
Tellurium hexafluoride	N/A		250	❑
Tetrafluoroethylene	10,000	❑	5,000	❑
Tetrafluorohydrazine	N/A		5,000	❑
Tetramethyllead	10,000	❑	1,000	❑
Tetramethylsilane	10,000	❑	10,000	❑
Tetranitromethane	10,000	❑	N/A	
Thionyl chloride	N/A		250	❑
Titanium tetrachloride	2,500	❑	N/A	
Toluene diisocyanate (unspecified isomer)	10,000	❑	N/A	
Toluene 2,6-diisocyanate	10,000	❑	N/A	
Toluene 2,4-diisocyanate	10,000	❑	N/A	
Trichloro (dichlorophenyl) silane	N/A		2,500	❑
Trichloro (chloromethyl) silane	N/A		100	❑
Trichlorosilane	10,000	❑	5,000	❑
Trifluoro-chloroethylene	10,000	❑	10,000	❑
Trimethylamine	10,000	❑	10,000	❑
Trimethylchlorosilane	10,000	❑	N/A	
Trimethyloxysilane	N/A		1,500	❑
Vinyl acetate monomer	15,000	❑	N/A	
Vinyl acetylene	10,000	❑	10,000	❑

Chemical Name	RMP Threshold Quantity (lbs)		PSM Threshold Quantity (lbs)	
Vinyl chloride	10,000	❑	10,000	❑
Vinyl ethyl ether	10,000	❑	10,000	❑
Vinyl fluoride	10,000	❑	10,000	❑
Vinyl methyl ether	10,000	❑	10,000	❑
Vinylidene chloride	10,000	❑	10,000	❑
Vinylidene fluoride	10,000	❑	10,000	❑

Appendix I

ERNS AND OTHER REPORTING FORMATS

ERNS INCIDENT NOTIFICATION REPORT

Regional Case Number: _____

| Reported (mm/dd/yy): | Time (HH/MM): | Multiple Report: ☐ | Regional Time (HH/MM): |

Recorded By: _____

Multiple Regional Case Number: _____

| Through NRC: ☐ | NRC Case Number: _____ | SSI Report: ☐ | CR Number: _____ |

A. REPORTER
Privacy Act

Confidentiality Requested: ☐   Reported By: _____

Organization Name: _____

Organization: (Check One) ☐ Discharger ☐ Public ☐ State ☐ Local ☐ Federal ☐ Unknown

*Address: _____   Phone: ( )   ext.:

City: _____   County: _____   State: _____   Zip: _____

B. DISCHARGER   Same As A ☐   Organization: (Check One) ☐ Private Co. ☐ Public ☐ State ☐ Local ☐ Federal ☐ Unknown

Discharger Name:	Phone: ( ) ext.:
Contact Name:	2nd Phone: ( ) ext.:
Address:	Facility ID Number:
City:	County:	State:	Zip:

C. INCIDENT LOCATION   Same As A ☐   Same As B ☐   Street or Approx. Location: _____

City: _____   County: _____   State: _____   Zip: _____   Milepost: _____

D. DATE   Discovery Date (mm/dd/yy): _____   Spill Date (mm/dd/yy): _____   Spill Time (HH/MM): _____

E. MATERIAL   Material Type: (Check One) ☐ Unknown ☐ Oil ☐ Haz Sub ☐ Other

	Material Name	CHRIS	CAS No.	UN DOT No.	Quantity	Units (Circle One)	Quantity In Water
1.						lb bbl drm unk / gal ton oth	
2.						lb bbl drm unk / gal ton oth	
3.						lb bbl drm unk / gal ton oth	

F. SOURCE   Source of Spill: (Check Any) ☐ Highway ☐ Railway ☐ Pipeline ☐ UST ☐ Fixed Facility ☐ Other
☐ Air Transport ☐ Vessel ☐ Offshore ☐ AST ☐ Unknown

Vehicle ID or Carrier No.: _____   Number of Tanks: _____   Tank Capacity: _____   Tank Units: (Circle One) lb bbl drm unk / gal ton oth

Source Description: _____

G. MEDIUM   Medium Affected: (Check Any) ☐ None ☐ Land ☐ Groundwater ☐ Other
☐ Air ☐ Water ☐ Within Facility ☐ Unknown

Waterway Affected: _____

H. CAUSE   Reported Cause: (Check Any) ☐ Transportation Accident ☐ Operational Error ☐ Dumping ☐ Other
☐ Equipment Failure ☐ Natural Phenomenon ☐ Unknown

Cause Description: _____

I. DAMAGE   No. of Injuries: _____ ☐ None   No. of Deaths: _____ ☐ None   Property Damage>$50,000: ☐

J. ACTIONS   Evacuation: ☐   Response Actions Taken: _____

K. NOTIFIED   Caller Has Notified: (Check Any) ☐ State/Local ☐ Discharger ☐ USCG ☐ Other ☐ Unknown

Agency Name: _____

L. COMMENTS   Comments: _____

Additional Information: (See Reverse Side) ☐

M. RESPONSE AND EVALUATION   Response Comments: _____

Agency Name: _____   (Check One) ☐ Local ☐ State ☐ Discharger ☐ Federal ☐ EPA ☐ Other ☐ Unknown

Agency Name: _____   (Check One) ☐ Local ☐ State ☐ Discharger ☐ Federal ☐ EPA ☐ Other ☐ Unknown

Agency Name: _____   (Check One) ☐ Local ☐ State ☐ Discharger ☐ Federal ☐ EPA ☐ Other ☐ Unknown

☐ =Critical Data For Data Quality   **CALLER INFORMATION**   Form Version 08/91

REGION 2 INCIDENT NOTIFICATION REPORT

Regional Case Number: _____

REGION-SPECIFIC	State Case #:		Drill: ☐
Caller Has Notified: (Check One)	☐ EPA ☐ NRC		Prevention Code(s):
Responsibility: (Check One)	☐ EPA ☐ USCG ☐ Other		
SPCC Letter: ☐	CWA 308 Spill Letter: ☐		Tributary To:
Referral:			

FOLLOW-UP	Update Date: (mm/dd/yy)	Updated By:

C. INCIDENT LOCATION	Dun & Bradstreet No.:

F. SOURCE	Source Code:

G. MEDIUM	Medium Code:	Threat Code(s):

H. CAUSE	Cause Code:

I. ACTIONS	No. of Persons Evacuated:

M. RESPONSE AND EVALUATION	Incident Status Code: (Check One)	☐ Classic Incident ☐ No Further Action	☐ Critical Incident ☐ Remedial Action	☐ Non-Critical Incident ☐ Field Simulation

Emergency Response Activity Within 24 Hours: ☐ Emergency Response Activity Date: (mm/dd/yy)

Responding OSC:

Action Memo Date: (mm/dd/yy)	Action Memo Approved: ☐	POLREP Date: (mm/dd/yy)
Release Investigation: ☐	On-Scene Monitoring: ☐	Telephone Assistance: ☐
TDD No.:	Enforcement Activities:	

Other Follow-up Comments:

☐ =Critical Data For Data Quality   **FOLLOW-UP INFORMATION**   Form Version 08/91

ARIP ACCIDENT REPORT FORM

OMB #: 2050-0065
Expires: September 30,1995

U.S. ENVIRONMENTAL PROTECTION AGENCY
ACCIDENTAL RELEASE INFORMATION PROGRAM

VERIFICATION AND INFORMATION SUPPLEMENT
INSTRUCTIONS

GENERAL INSTRUCTIONS

The Accidental Release Information Program (ARIP) is administered by the U.S. Environmental Protection Agency (EPA). The purpose of this program is to learn more about the causes and consequences of accidental releases of hazardous substances from fixed facilities and the actions that have been or could have been effective in preventing them from occurring. The collected information will serve to support a range of chemical accident prevention and preparedness efforts involving industry, local and state government, and EPA regions and headquarters.

Please read the instructions before you verify the information or answer the questions. If you need further assistance, please contact the person identified in the cover letter.

ORGANIZATION

There are two parts to this survey. Both parts must be completed and all questions must be answered.

PART A.   Emergency Response Notification System - Verification

The Emergency Response Notification System (ERNS) is a national computer database and retrieval system that is used to store information on releases of oil and hazardous substances. ERNS provides a mechanism for documenting and verifying incident notification information as initially reported to the National Response Center (NRC), EPA, and/or the U.S. Coast Guard. This part of the survey includes an ERNS printout of available information pertaining your particular release event. It is important that you verify this information, make corrections as needed, and provide any missing information.

Please note the identification numbers for your reported event at the top of this section. These numbers are the ERNS database number and/or the NRC report number. Make sure that you **IDENTIFY ALL PAGES WITH AT LEAST ONE OF THESE NUMBERS.**

PART B.   Accidental Release Prevention - Supplemental Information

This part of the survey questionnaire requests information to supplement reports you may have submitted to the National Response Center (NRC) and other federal, state, or local authorities. The questionnaire is divided into three sections:

Section I - Facility Profile

This section asks several questions about your facility, e.g., location, product, and current status of operations.

Section II - Hazardous Substance Release Profile

This section asks several key questions concerning the reported release.
It is important that you respond as accurately as you can based on the

information available to you. If more space or comments to clarify your
response are needed, please use additional pages.

Section III - Prevention Profile

This section asks you to provide an assessment of prevention plans and
technologies at your facility and any changes that will be initiated
because of the release incident. Additional space for alternate answers
and/or details is provided. Please attach additional pages if
necessary.

AGENCY DISCLOSURE OF ESTIMATED BURDEN

Public reporting burden for this collection of information is estimated
to average 24.5 hours, including time for reviewing instructions, searching
existing data sources, gathering and maintaining data needed, and completing
and reviewing the collection of information. Send comments regarding the
estimated burden or any other aspect of this collection of information,
including suggestions for reducing the burden, to Director, Regulatory
Information Division, Mail Code 2136, U.S. Environmental Protection Agency,
401 M St., S.W., Washington, D.C. 20460; and to Paperwork Reduction Project
(OMB # 2050-0065), Office of Information and Regulatory Affairs, Office of
Management and Budget, Washington, D.C. 20503.

DEFINITIONS

*Please refer to the definitions below to clarify the precise meaning and
use of the terms in the questionnaire.*

By-pass: A piping system designed to provide an alternate pathway for gas or
liquid streams that detours around a normal pathway. A by-pass condition
refers to a system's operation using available bypass systems. Certain
instrument control alarms and interlocks may also be "by-passed" during
abnormal operating conditions.

Cause-Consequence Analysis[1]: A diagram display of the interrelationships
between accident outcomes and their basic causes. This analysis is a blend of
the Fault Tree and Event Tree analysis.

Containment System: Dikes, curbs, vaults, ponds, and the like that serve to
collect and temporarily hold spilled materials until such time as they are
removed, disposed of, or transferred to a secure storage vessel.

Dow and Mond Indices[1]: A method for relative ranking of the risks at a
chemical process plant. This method assigns penalties to process materials
and conditions that can contribute to an accident. Credits are assigned to
plant safety procedures that can mitigate the effects of an accident.

Equipment (Mechanical) Failure: Failure of process or storage vessels,
valves, piping, pumps or other equipment connecting vessels in a process that
allows a loss of containment.

Event Tree Analysis[1]: Considers operator response or safety system response to an initiating event in determining accident outcome. This analysis results in accident sequences.

Facility Boundary: Fence line or property line marking the perimeter of a facility.

Failure Modes/Effects Analysis[1]: A method for tabulating the system/plant equipment and their respective failure modes (description of how the equipment or system fails). The tabulation includes the effects of each failure mode on the system/plant and a critical ranking of them.

Fault Tree Analysis[1]: A deductive technique that focuses on determining the causes of one particular accident event. The causes are determined using the fault tree - a graphic model that displays the various combinations of equipment faults and failures that can result in an accident event.

Federal Authority: Any federal government official delegated the responsibility under the Superfund statute for activities related to hazardous substance releases (e.g., National Response Center, U.S. Environmental Protection Agency and its regional offices).

General Public: Persons not present within the facility boundaries at the time the release occurred and/or with no business association to the facility owner (e.g., residents near the facility).

Hazard Assessment[1]: Formal procedures employed to identify potential risks that could lead to an accidental release (e.g., Fault Tree analysis).

Hazard and Operability Studies (HAZOP)[1]: Formal team brainstorming to systematically identify hazards and operability problems throughout an entire facility. Certain guide-words such as "no flow" and "no cooling" are used. The consequences of credible deviations associated with the guide-words are identified and assessed.

Hazardous Substance: Any element, compound, mixture, solution, or substance designated under section 102 of the Comprehensive Environmental Response, Compensation, and Liability Act (CERCLA) or section 3001 of the Solid Waste Disposal Act.

Human Error Analysis[1] (also know as Human Factors Analysis): A systematic evaluation of the factors that influence the performance, procedures, and techniques of human operators, maintenance staff, and other personnel. It will identify errors and likely situations that can cause an accident.

Immediate Response[1]: Application of equipment, systems, and procedures to capture, neutralize, or destroy a hazardous substance <u>before</u> it is released to the environment (e.g., scrubber).

Local Authority: Any local government official responsible for remedial or related activities connected with a hazardous substance release (e.g., Local Emergency Response Committee (LEPC), fire department).

Loss of Containment: Accidental release of hazardous substances from a process or storage vessel, interconnecting equipment, and/or control equipment to the environment.

Migration: The movement of a substance from one place to another in air, water, soil, or other media.

Operator Error: A mistake (e.g., leaving a valve open, failure to respond to process alarms, failure to maintain process variables or conditions at set point) made during operation of a process by the operator resulting in a release or loss of containment.

Owner: The legally designated individual, partnership, or parties that own the facility.

POTW: Publicly Owned Treatment Works.

Probabilistic Risk Assessment[1]: The overall measure of risk determined through numerical evaluation of both accidental consequences and probabilities. This method is used to assess comparative risk where alternative designs exist.

Process Control and Monitoring[1]: Control and detection equipments that provide information on the process status, standard operating conditions or parameters, and possible or imminent releases (e.g., pressure sensors, temperature sensors, chemical detectors on process lines).

Process Design[1]: Design of process equipment and systems to limit the potential for accidental releases (e.g., redundant systems).

Process Vessel: A tank, reactor, vat, or other piece of equipment in which substances are blended to form a mixture or are reacted to convert them to some other product or form.

Release: Any unintentional or accidental spilling, leaking, flowing, pumping, pouring, emitting, emptying, discharging, injecting, escaping, leaching, dumping, or disposing of a hazardous substance into the environment from a storage or process vessel.

Responding Official: Person responsible for the final review of the information provided in the survey questionnaire for completeness and accuracy (e.g., facility safety officer, environmental engineer, plant manager).

Response[1]: Application of equipment, systems, and procedures to capture, neutralize, or destroy a hazardous substance <u>after</u> it is released to the environment (e.g., cleanup).

Standard Industrial Classification: The federal government categories of business activity. See Standard Industrial Classification Manual, Office of Management and Budget, U.S. Government Printing Office, Washington, D.C.

State Authority: Any state government official responsible for remedial or related activities connected with a hazardous substance release (e.g., State Emergency Response Commission (SERC), state transportation office).

Storage Vessel: Any container (e.g., tank, drum, bottle, tank car, cylinder) used to hold a raw or input material, a product, or a by-product at ambient conditions or at an elevated or reduced temperature or pressure.

Upset: Process deviation from standard conditions because of a malfunction or failure of process controls, alarms, or backup systems. These conditions could result from operator error, mechanical or equipment failure, or from unexpected events such as fire, explosion, power loss, or water loss.

What If Analysis[1]: Considers consequences associated with events that occur as a result of failures involving equipment, design, or procedures. All possible system failures are collected in a list and evaluated (e.g., "what if the feed pump fails"). This method requires a basic understanding of what is intended and the ability to combine possible deviations and to reject incredible situations.

1. Definition derived from <u>Guidelines for Hazard Evaluation Procedures</u>, AICHE, 1985, and from the <u>Review of Emergency Systems</u>, EPA, June, 1988.

OMB #: 2050-0065
Expires:

U.S. ENVIRONMENTAL PROTECTION AGENCY
ACCIDENTAL RELEASE INFORMATION PROGRAM

PART A.   **EMERGENCY RESPONSE NOTIFICATION SYSTEM - VERIFICATION**

Information regarding an accidental release incident in your facility has been recorded in the Emergency Response Notification System (ERNS). Below is the information available in ERNS regarding this release. Please verify the information by making any corrections and/or by providing any missing information in the spaces provided (attach additional pages as necessary).

1. Facility: _____

Dun & Bradstreet Number _____-_____-_____

Street   _____
City   _____
County   _____
State   _____   Zip _____

Telephone (   ) _____

2. Spill Location: (____ Check here if same as Facility Address)

Street   _____
City   _____
County   _____
State   _____   Zip _____

Telephone (   ) _____

Latitude (Deg/Min) _____/_____   Longitude _____/_____

3. Primary Chemical Released: _____

4. ERNS Reporting Date/Time: _____/_____
                                  (mm/dd/yy)   (24-hr clock)

5. Reported through NRC?   Yes_____   No_____

ERNS/REGIONAL CASE # _____ **/ NRC #** _____

6. Federal, State, and Local Authorities Notified:

(e.g., NRC, EPA Regional Office, SERC, DNR, LEPC, Police, and others. Show dates as mm/dd/yy; times in 24 hour clock.)

Agency	Date	Time	Person Contacted

7. Responding Agencies:

8. Response Action(s):

ERNS/REGIONAL CASE # _____ **/ NRC #** _____

PART B.   ACCIDENTAL RELEASE PREVENTION - SUPPLEMENTAL INFORMATION

<u>SECTION I.   FACILITY PROFILE</u>

1.   **Plant Manager/Facility Owner:**_____

2.   **Responding Official:**   _____

                Title: _____

              Address: _____
                                   _____
                                   _____

              Telephone: (   )   _____

    Signature: _____   **Date:**_____

3.   Please provide the four-digit Standard Industrial Classification (SIC)
    codes that best describe your facility operations:

    SIC code(s):   _____,   _____,   _____
                  (Primary)

    Primary product or service: _____
    _____
    _____

4.   Indicate the total number of employees typically at the facility
    (include all full-time and part time employees, all employees on sick
    leave, paid holidays, paid vacations, managers and corporate officers at
    the facility, and contractors):

        Number of Employees: _____

ERNS/REGIONAL CASE # _____ / NRC # _____

SECTION II. HAZARDOUS SUBSTANCE RELEASE PROFILE

For the following section, if exact responses cannot be provided please provide estimates using your best professional judgment.

5. **Date/Time Release Began:** _____; _____

                                         (month/day/year)   (24-hr clock)

                Ended: _____; _____

                                         (month/day/year)   (24-hr clock)

6. In the table below, provide release estimates for the primary chemical released (in lbs, only) to each media. Quantities released to each media should add up to the total quantity released. For solutions, adjust the quantity of the chemical released for chemical concentration (e.g., report 1,000 lbs of 50% sulfuric acid released as 500 lbs sulfuric acid). For multiple chemicals attach additional pages as necessary.

Chemical Name: _____

CAS Number: _____

Concentration (wt%): _____

Physical State at time of release: _____

Released To: **Quantity (lbs):**

        Air   _____

        Surface Water   _____

        Land   _____

        Treatment Facility   _____

Total Quantity Released: _____

7. Check the item below that best describes when the release occurred:

    a. ___ During routine operation

    b. ___ During routine startup

    c. ___ While in process of shutting down operations

    d. ___ While unit was shutdown for maintenance/product changeover, etc.

    e. ___ During special test, or non-standard, trial run conditions

    f. ___ During startup of new construction, new equipment

    g. ___ Other (please describe):

_____
_____
_____

ERNS/REGIONAL CASE # _____ **/ NRC #** _____

8. Check the item below that best describes the status of the facility,
    unit, or process line as a result of the release:

    a. ____ No interruption; continued operations
    b. ____ Restarted after release
    c. ____ Shut down for repairs; with plans to restart
    d. ____ Permanently closed
    e. ____ Other (please describe):

    _____
    _____
    _____

9. Check the one item below that best describes the location of the loss of
    containment in the specified area:

    a. **Process Vessel:** ___ wall, ___ overflow, ___ vent, ___ drain
    b. **Storage vessel:** ___ wall, ___ overflow, ___ vent, ___ drain
    c. **Valve:** ___ flange, ___ seal, ___ body
    d. **Piping:** ___ flange, ___ joint, ___ elbow, ___ wall
    e. **Pump:** ___ flange, ___ seal, ___ body
    f. Other process equipment (please describe):

    _____
    _____
    _____

10. How was the release first discovered? (check as many as apply)

    a. ____ Process control device indication
    b. ____ Chemical specific detector, alarm
    c. ____ Observation by employee(s)
    d. ____ Explosion/fire
    e. ____ Third party notification
    f. ____ Other (please describe):

    _____
    _____
    _____

11. Check one item below that best describes what initiated the release:

    a. ____ Equipment failure
    b. ____ Operator error

ERNS/REGIONAL CASE # _____ / NRC # _____

12. Indicate other factors that contributed to the equipment failure or operator error (check as many as apply and elaborate below):

   a. ____ "Upset" condition
   b. ____ "By-pass" condition
   c. ____ Maintenance activity
   d. ____ Training deficiencies
   e. ____ Inappropriate operating procedures
   f. ____ Faulty process design
   g. ____ Unsuitable equipment
   h. ____ Unusual weather Conditions
   i. ____ Other (please describe):

   _____
   _____
   _____

13. Provide a brief chronological description of the events that led up to and contributed to the release event (if helpful, include a sketch). Briefly discuss the results of your investigation. Use additional pages as necessary.

   _____
   _____
   _____
   _____
   _____
   _____
   _____
   _____
   _____
   _____
   _____
   _____

14. Check all items that describe the end effects of the release event:

   a. ____ Spill
   b. ____ Vapor release
   c. ____ Explosion
   d. ____ Fire
   e. ____ Other (describe):

   _____
   _____
   _____

ERNS/REGIONAL CASE # _____ **/ NRC #** _____

15. Was the general public notified? **Yes** _____ **No** _____

    If yes, indicate the type of communication technologies used to alert
    and notify the public to evacuate or take other safety measures. Check
    as many items as apply:

    a. ____ Door-to-door notification
    b. ____ Loudspeakers/public access system
    c. ____ Tone alert radio/pagers
    d. ____ Siren/alarms
    e. ____ Modulated power lines
    f. ____ Aircraft
    g. ____ Radio
    h. ____ Television
    i. ____ Cable override
    j. ____ Telephone
    k. ____ Other (please describe):

    _____
    _____
    _____

16. Indicate the number of persons injured, hospitalized (as opposed to
    treated and released) and fatalities that occurred as a result of the
    release (indicate with NA if not known):

	Injuries	**Hospitalized**	**Fatalities**
Facility employees	_____	_____	_____
Contractors	_____	_____	_____
General public	_____	_____	_____
Responders	_____	_____	_____

17. Indicate the number of persons evacuated and/or sheltered-in-place as a
    result of the release (indicate with NA if not known):

	Evacuated	**Sheltered in Place**
Facility employees	_____	_____
Contractors	_____	_____
General public	_____	_____

ERNS/REGIONAL CASE # _____ **/ NRC #** _____

18. Describe the immediate response activities taken to mitigate the release (capture, neutralize or destroy a toxic chemical before it is released into the environment). Check as many as apply.

   a. ____ Reduce system pressure/temperature
   b. ____ Apply spray scrubber/curtain
   c. ____ Transfer contents from failed equipment
   d. ____ Dilute and/or neutralize
   e. ____ Containment
   f. ____ Plant/process shutdown
   g. ____ Divert release to treatment
   h. ____ Vacuum/release recovery
   i. ____ Incineration/flares
   j. ____ None
   k. ____ Other (describe):

   _____
   _____
   _____

19. Indicate the environmental effects that occurred as a result of the release:

   a. ____ Fish Kills
   b. ____ Vegetation damage
   c. ____ Soil contamination
   d. ____ Groundwater contamination
   e. ____ Wildlife kills
   f. ____ None
   g. ____ Other (please describe):

   _____
   _____
   _____

20. Estimate the financial impact of the accidental release for the facility (e.g., cleanup cost, outside contractors cost, hours/wages diverted to cleanup or lost to shutdown, loss of production) and for the general public (e.g., damage to natural resources, public and private properties). An aggregate figure may be provided if a breakdown is not available.

   a. Facility Costs: $_____
   b. General Public Costs: $_____

   Total Costs: $_____

ERNS/REGIONAL CASE # _____ / NRC # _____

SECTION III. PREVENTION PROFILE

21a. What formalized hazard evaluation was performed prior to this release at
the process or storage area within your facility where the accident
occurred? When was it last conducted? How frequently is this
evaluation conducted (e.g. every 2 years)? Indicate frequency in years
and date last conducted as mm/dd/yy.

		Frequency	Last Conducted
a. ___	Cause-Consequence analyses	_____	_____
b. ___	Dow and Mond Hazard Indices	_____	_____
c. ___	Event Tree analyses	_____	_____
d. ___	Failure Modes/Effects analyses	_____	_____
e. ___	Fault Tree analyses	_____	_____
f. ___	HAZOP Studies	_____	_____
g. ___	Human Error analyses	_____	_____
h. ___	Probabilistic Risk Assessments	_____	_____
i. ___	What If analyses	_____	_____
j. ___	No evaluation ever done for this area		
k. ___	Other evaluation (describe, indicate frequency, date done):		

_____
_____
_____
_____
_____

21b. Was the hazard evaluation performed effective in predicting this release
event? Why or why not?

_____
_____
_____
_____
_____
_____

ERNS/REGIONAL CASE # _____ **/ NRC #** _____

22a. Identify the training, procedures, and/or management practices used at this facility prior to this release to prevent accidental releases. Check all that apply.

- **a.** ___ Preventive Maintenance/Inspections
- **b.** ___ Accident Investigations
- **c.** ___ Audits
- **d.** ___ Inventory/capacity reductions
- **e.** ___ Employee safety training
- **f.** ___ Standard operating procedures
- **g.** ___ Emergency response training
- **h.** ___ None
- **i.** ___ Other (please describe):

_____
_____
_____
_____
_____

22b. Describe any changes to existing training, procedures and management practices, or what new types of training, procedures and management practices are or will be implemented as a result of this release?

_____
_____
_____
_____
_____
_____

23a. What engineering systems or controls were in use prior to the release at the process or storage area within your facility where the accident occurred? Check all that apply.

- **a.** ___ Backup/Redundant systems
- **b.** ___ Automatic Shut-offs
- **c.** ___ Bypass/Surge systems
- **d.** ___ Manual Overrides
- **e.** ___ Controls for operations monitoring and warning
- **f.** ___ Interlocks
- **g.** ___ None
- **h.** ___ Other (please describe):

_____
_____
_____
_____

ERNS/REGIONAL CASE # _____ **/ NRC #** _____

23b. Describe any changes to the existing engineering systems or controls, and any new types of engineering systems/controls that are or will be implemented as a result of this release:

_____
_____
_____
_____
_____
_____
_____
_____
_____

HMIRS ACCIDENT REPORT FORM

DEPARTMENT OF TRANSPORTATION
HAZARDOUS MATERIALS INCIDENT REPORT

Form Approved OMB No. 2137-0039

INSTRUCTIONS: Submit this report in duplicate to the Information Systems Manager, Office of Hazardous Materials Transportation, DHM-63, Research and Special Programs Administration, U.S. Department of Transportation, Washington, D.C. 20590. If space provided for any item is inadequate, complete that item under Section IX, keying to the entry number being completed. Copies of this form, in limited quantities, may be obtained from the Information Systems Manager, Office of Hazardous Materials Transportation. Additional copies in this prescribed format may be reproduced and used, if on the same size and kind of paper.

I. MODE, DATE, AND LOCATION OF INCIDENT

1. MODE OF TRANSPORTATION: ☐ AIR ☒ HIGHWAY ☐ RAIL ☐ WATER ☐ OTHER ....

2. DATE AND TIME OF INCIDENT (Use Military Time, e.g. 8:30am = 0830. noon = 1200, 6pm = 1800, midnight = 2400).
Date: 3 / 7 / 1995   TIME: 1100 a.m.

3. LOCATION OF INCIDENT (Include airport name in ROUTE/STREET if incident occurs at an airport.)
CITY: Belvoir   STATE: Virginia
COUNTY: Fairfax   ROUTE/STREET: US Route 1

II. DESCRIPTION OF CARRIER, COMPANY, OR INDIVIDUAL REPORTING

4. FULL NAME
ABC Trucking Company

5. ADDRESS (Principal place of business)
1492 Columbus Avenue
Richmond, VA 23021

6. LIST YOUR OMC MOTOR CARRIER CENSUS NUMBER, REPORTING RAILROAD ALPHABETIC CODE, MERCHANT VESSEL NAME AND ID NUMBER OR OTHER REPORTING CODE OR NUMBER. MC 654321

III. SHIPMENT INFORMATION (From Shipping Paper or Packaging)

7. SHIPPER NAME AND ADDRESS (Principal place of business)
Scientific Division – American Hotel Supply
1101 South Peachtree Street
Atlanta, GA 30303

8. CONSIGNEE NAME AND ADDRESS (Principal place of business)
J & J Chemicals
9801 Sluice Parkway
Newark, NJ 07101

9. ORIGIN ADDRESS (If different from Shipper address)
N/A

10. DESTINATION ADDRESS (If different from Consignee address)
1506 Wayne Street
Alexandria, VA 22301

11. SHIPPING PAPER/WAYBILL IDENTIFICATION NO. Carrier's PRO 98765

IV. HAZARDOUS MATERIAL(S) SPILLED (NOTE: REFERENCE 49 CFR SECTION 172.101.)

12. PROPER SHIPPING NAME	13. CHEMICAL/TRADE NAME	14. HAZARD CLASS	15. IDENTIFICATION NUMBER (e.g. UN 2784, NA 2020)
Acetone	N/A	Flammable Liquid	UN 1090

16. IS MATERIAL A HAZARDOUS SUBSTANCE? ☒ YES ☐ NO   **17. WAS THE RQ MET?** ☐ YES ☒ NO

V. CONSEQUENCES OF INCIDENT, DUE TO THE HAZARDOUS MATERIAL

18. ESTIMATED QUANTITY HAZARDOUS MATERIAL RELEASED (Include units of measurement)	19. FATALITIES	20. HOSPITALIZED INJURIES	21. NON-HOSPITALIZED INJURIES
45 Gallons	None	None	1

22. NUMBER OF PEOPLE EVACUATED None

23. ESTIMATED DOLLAR AMOUNT OF LOSS AND/OR PROPERTY DAMAGE, INCLUDING COST OF DECONTAMINATION OR CLEANUP (Round off in dollars)

A. PRODUCT LOSS	B. CARRIER DAMAGE	C. PUBLIC/PRIVATE PROPERTY DAMAGE	D. DECONTAMINATION/ CLEANUP	E. OTHER
$90.00	N/A	N/A	$100.00	N/A

24. CONSEQUENCES ASSOCIATED WITH THE INCIDENT: ☒ SPILLAGE ☐ FIRE ☐ EXPLOSION ☐ VAPOR (GAS) DISPERSION ☐ ENVIRONMENTAL DAMAGE ☐ MATERIAL ENTERED WATERWAY/SEWER ☐ NONE ☐ OTHER: _____

VI. TRANSPORT ENVIRONMENT

25. INDICATE TYPE(S) OF VEHICLE(S) INVOLVED ☐ CARGO TANK ☒ VAN TRUCK/TRAILER ☐ FLAT BED TRUCK/TRAILER ☐ TANK CAR ☐ RAIL CAR ☐ TOFC/COFC ☐ AIRCRAFT ☐ BARGE ☐ SHIP ☐ OTHER: _____

26. TRANSPORTATION PHASE DURING WHICH INCIDENT OCCURRED OR WAS DISCOVERED: ☒ EN ROUTE BETWEEN ORIGIN/DESTINATION ☐ LOADING ☐ UNLOADING ☐ TEMPORARY STORAGE/TERMINAL

27. LAND USE AT INCIDENT SITE ☐ INDUSTRIAL ☒ COMMERCIAL ☐ RESIDENTIAL ☐ AGRICULTURAL ☐ UNDEVELOPED

28. COMMUNITY TYPE AT SITE ☐ URBAN ☒ SUBURBAN ☐ RURAL

29. WAS THE SPILL THE RESULT OF A VEHICLE ACCIDENT/DERAILMENT? ☒ YES ☐ NO
IF YES AND APPLICABLE ANSWER PARTS A THRU C

A. ESTIMATED SPEED	B. HIGHWAY TYPE	C. TOTAL NUMBER OF LANES	SPACE FOR DOT USE ONLY
25 mph	☒ DIVIDED/LIMITED ACCESS ☐ UNDIVIDED	☐ ONE ☐ THREE ☒ TWO ☐ FOUR OR MORE	

FORM DOT F 5800.1 (Rev 6/89) Supersedes DOT F 5800.1 (10/70) (9/1/79)   THIS FORM MAY BE REPRODUCED

VII. PACKAGING INFORMATION: If the package is overpacked (consists of several packages, e.g. glass jars within a fiberboard box), begin with Column A for information on the innermost package.

ITEM	A (Inner)	B (Outer)	C
30. TYPE OF PACKAGING, INCLUDING INNER RECEPTACLES (e.g. Steel drum, tank can)	Plastic Liner	Steel Drum	
31. CAPACITY OR WEIGHT PER UNIT PACKAGE (e.g. 55 gallons, 65 lbs.)	55 Gallons	55 Gallons	
32. NUMBER OF PACKAGES OF SAME TYPE WHICH FAILED IN IDENTICAL MANNER	1	1	
33. NUMBER OF PACKAGES OF SAME TYPE IN SHIPMENT	12	12	
34. PACKAGE SPECIFICATION IDENTIFICATION (e.g. DOT 17E, DOT 105A100, UN 1A1 or none)	DOT 2SL	DOT 17H	
35. ANY OTHER PACKAGING MARKINGS (e.g. STC. 18/16-55-88, Y1.4/150/87)	55-12-93	STC 18/16-55-92	
36. NAME AND ADDRESS, SYMBOL OR REGISTRATION NUMBER OF PACKAGING MANUFACTURER	AAA - Toledo, OH	FUBAR - Flint, MI	
37. SERIAL NUMBER OF CYLINDERS, PORTABLE TANKS, CARGO TANKS, TANK CARS	N/A	N/A	
38. TYPE OF LABELING OR PLACARDING APPLIED	None	Flammable Liquid	
39. IF RECONDITIONED OR REQUALIFIED — A. REGISTRATION NUMBER OR SYMBOL	N/A	DOT 51000	
39. IF RECONDITIONED OR REQUALIFIED — B. DATE OF LAST TEST OR INSPECTION	N/A	2/94	
40. EXEMPTION/APPROVAL/COMPETENT AUTHORITY NUMBER, IF APPLICABLE (e.g. DOT E1012)	N/A	N/A	

VIII. DESCRIPTION OF PACKAGING FAILURE: Check all applicable boxes for the package(s) identified above.

41. ACTION CONTRIBUTING TO PACKAGING FAILURE

	A	B	C			A	B	C	
a.	☒	☒	☐	TRANSPORT VEHICLE COLLISION	j.	☒	☒	☐	CORROSION
b.	☐	☐	☐	TRANSPORT VEHICLE OVERTURN	k.	☐	☐	☐	METAL FATIGUE
c.	☐	☐	☐	OVERLOADING/OVERFILLING	l.	☐	☐	☐	FRICTION/RUBBING
d.	☐	☐	☐	LOOSE FITTINGS, VALVES	m.	☐	☐	☐	FIRE/HEAT
e.	☐	☐	☐	DEFECTIVE FITTINGS, VALVES	n.	☐	☐	☐	FREEZING
f.	☐	☐	☐	DROPPED	o.	☐	☐	☐	VENTING
g.	☒	☒	☐	STRUCK/RAMMED	p.	☐	☐	☐	VANDALISM
h.	☐	☐	☐	IMPROPER LOADING	q.	☐	☐	☐	INCOMPATIBLE MATERIALS
i.	☐	☐	☒	IMPROPER BLOCKING	r.	☐	☐	☐	OTHER _____

42. OBJECT CAUSING FAILURE

	A	B	C	
a.	☒	☒	☐	OTHER FREIGHT
b.	☐	☐	☐	FORKLIFT
c.	☐	☐	☐	NAIL/PROTRUSION
d.	☐	☐	☐	OTHER TRANSPORT VEHICLE
e.	☐	☐	☐	WATER/OTHER LIQUID
f.	☐	☐	☐	GROUND/FLOOR/ROADWAY
g.	☐	☐	☐	ROADSIDE OBSTACLE
h.	☐	☐	☐	NONE
i.	☐	☐	☐	OTHER _____

43. HOW PACKAGE(S) FAILED

	A	B	C	
a.	☒	☒	☐	PUNCTURED
b.	☐	☐	☐	CRACKED
c.	☐	☐	☐	BURST/INTERNAL PRESSURE
d.	☐	☐	☐	RIPPED
e.	☐	☒	☐	CRUSHED
f.	☐	☐	☐	RUBBED/ABRADED
g.	☐	☐	☐	RUPTURED
h.	☐	☐	☐	OTHER _____

44. PACKAGE AREA THAT FAILED

	A	B	C	
a.	☐	☐	☐	END, FORWARD
b.	☐	☐	☐	END, REAR
c.	☐	☐	☐	SIDE, RIGHT
d.	☐	☐	☐	SIDE, LEFT
e.	☐	☐	☐	TOP
f.	☐	☐	☐	BOTTOM
g.	☒	☒	☐	CENTER
h.	☐	☐	☐	OTHER _____

45. WHAT FAILED ON PACKAGE(S)

	A	B	C	
a.	☒	☒	☐	BASIC PACKAGE MATERIAL
b.	☐	☐	☐	FITTING/VALVE
c.	☐	☐	☐	CLOSURE
d.	☐	☒	☐	CHIME
e.	☐	☐	☐	WELD/SEAM
f.	☐	☐	☐	HOSE/PIPING
g.	☐	☒	☐	INNER LINER
h.	☐	☐	☐	OTHER _____

IX. DESCRIPTION OF EVENTS: Describe the sequence of events that led to incident, action taken at time discovered, and action taken to prevent future incidents. Include any recommendations to improve packaging, handling, or transportation of hazardous materials. Photographs and diagrams should be submitted when necessary for clarification. ATTACH A COPY OF THE HAZARDOUS WASTE MANIFEST FOR INCIDENTS INVOLVING HAZARDOUS WASTE. Continue on additional sheets if necessary.

Our vehicle was involved in a minor traffic accident which caused the load to shift and puncture one of the drums. The leaking drum and all of the spilled Acetone was removed for disposal by Hazmat Cleanup Service Inc. to their site at 9987 Old Town Road, March, VA. The vehicle was taken to our Alexandria terminal and cleaned (washed and steamed).

A Highway Patrolman on the scene had some of the spilled liquid splash on his hand. He received first aid at the scene for his skin irritation.

46. NAME OF PERSON RESPONSIBLE FOR PREPARING REPORT	47. SIGNATURE	
Gonzalez		
48. TITLE OF PERSON RESPONSIBLE FOR PREPARING REPORT	49. TELEPHONE NUMBER (Area Code)	50. DATE REPORT SIGNED
Traffic Safety Coordinator	(703) 666-4321	April 5, 1995

HLPAD ACCIDENT REPORT FORM

OMB No. 2137-0047

ACCIDENT REPORT-HAZARDOUS LIQUID PIPELINE

Report Date

No. _7000-1_
(DOT)

PART A—OPERATOR INFORMATION

1.) Name of operator _First Pipeline Company_

2.) Principal business address _One Tenley Place, 6100 So. Yale P. O._
Anytown, PA

(city)   *(state)*   *(zip code)*

3.) Is pipeline interstate? ☒ yes   ☐ no

PART B—TIME AND LOCATION OF ACCIDENT

1.) Date: *(month)* May   *(day)* 9   *(year)* 1995

2.) Hour _(24 hour clock)_ 2010

3.) If onshore give state *(including Puerto Rico and Washington, D.C.).*
and county or city. _Jefferson Co., Anytown, Pa._

4.) If offshore, give offshore coordinates _____

5.) Did accident occur on Federal Land? ☐ yes   ☒ no
(See instructions for definition of Federal Land.)

6.) Specific location *(If location is near offshore platforms, buildings, or other landmarks, such as highways, waterways, or railroads, attach a sketch or drawing showing relationship of accident location to these landmarks)*
3 miles West of Main St. — Thence 3 miles South of HWY 9 on Coker Road,
thence 300' West of Coker Road at Dance Creek.

PART C—ORIGIN OF RELEASE OF LIQUID OR VAPOR.   *(Check all applicable items)*

1.) Part of system involved:
☒ line pipe   ☐ tank farm   ☐ pump station

2.) Item involved: ☒ pipe   ☐ valve   ☐ scraper trap   ☐ pump
☐ welding fitting   ☐ girth weld   ☐ tank
☐ bolted fitting   ☐ longitudial weld

Other *(specify)* _____

3.) Year item installed _1949_

PART D—CAUSE OF ACCIDENT

☐ corrosion   ☐ failed weld   ☐ incorrect operation by operator personnel
☐ failed pipe   ☐ outside force damage
☐ malfunction of control or relief equipment.
☒ other *(specify)* _Pipeline was punctured by rifle bullet._

PART E—DEATH OR INJURY

1.) Number of persons killed. _0_
_____ Operator employees _____ Non-employees

2.) Number of persons injured. _0_
_____ Operator employees _____ Non-employees

PART F—ESTIMATED TOTAL PROPERTY DAMAGE
$ _25,000._

PART G—COMMODITY SPILLED

1.) Name of commodity spilled: _Gasoline_

2.) Classification of commodity spilled:
☐ Petroleum   Petroleum product ☐ HVL or ☒ Non-HVL

3.) Estimated amount of commodity involved
_500_ Barrels spilled _2_ Barrels recovered

4.) Was there an explosion?
☐ yes   ☒ no

5.) Was there a Fire?
☐ yes   ☒ no

```
 ┌──────────────────────────────────────────────────────────────────────┐
 │    INSTRUCTIONS: Answer sections H, I, or J only if it applies to the   │
 │                  particular accident being reported.                    │
 ├──────────────────────────────────────────────────────────────────────┤
```

PART H—OCCURRED IN LINE PIPE

1.) Nominal diameter *(inches)* __8"__   2.) Wall thickness *(inches)* _0.322_

3.) SMYS *(psi)* _35,000_ 4.) Type of joint: ☒ welded   ☐ flanged   ☐ threaded   ☐ coupled   ☐ other

5.) Pipe was ☐ Below ground   ☒ Above ground

6.) Maximum operating pressure *(psig)* _700_

7. Pressure at time and location of accident *(psig)* _4.5_

8.) Had there been a pressure test on system?
   ☐ yes   ☒ no

9.) Duration of test *(hrs)* _____

10.) Maximum test pressure *(psig)* _____

11.) Date of latest test _____

PART I—CAUSED BY CORROSION

1. Location of corrosion
   ☐ internal   ☐ external

2. Facility coated?
   ☐ yes   ☐ no

3. Facility under cathodic protection?
   ☐ yes   ☐ no

4. Type of corrosion
   ☐ galvanic   ☐ other *(Specify)*
   _____

PART J—CAUSED BY OUTSIDE FORCE

1. ☐ Damage by operator or its contractor
   ☒ Damage by others
   ☐ Damage by natural forces
   ☐ Landslide
   ☐ Subsidence
   ☐ Washout
   ☐ Frostheave
   ☐ Earthquake
   ☐ Ship anchor
   ☐ Mudslide
   ☐ Fishing Operations
   Other _Rifle Bullet_

2. Was a damage prevention program in effect
   ☒ yes   ☐ no
3. If yes, was the program
   ☒ "one-call"   ☐ other _____
4. Did excavator call?
   ☐ yes   ☐ no   N/A
5. Was pipeline location temporarily marked for the excavator?
   ☐ yes   ☐ no   N/A

PART K—ACCOUNT OF ACCIDENT

5/9/95

2010	Police Dept. Reported leak to us.
2015	Pressure loss indicated that we had leak.
2044	One mile area around leak site evacuated.
2100	Damaged pipe was under water because of heavy rain.
2117	Valve on North side of leak site closed.
2130	Valve on South side of leak site closed.

5/9/95

0400	Water receded and wooden peg was driven into bullet hole.
0645	Installed temporary repair clamp.

NAME AND TITLE OF OPERATOR OFFICIAL FILING THIS REPORT.
John Smith, Manager, Pipeline Operations

_(999) 123-4567_   _May 10, 1995_
Telephone no. *(Including area code)*   Date

DOT Form 7000-1 (4-85)

IMIS ACCIDENT REPORT FORM

•••••••••••••••••••••••••••

Activity Number: 109294944 SIC 2421   Open Date: 2/11/92 NONUNION
**** ACCIDENT DATA ••••
    SUMMARY # 793224 DATE: 2-11-92
DESCRIP:CAUGHT BETWEEN STATIONARY BEAM AND MOVING MACHINE PART
    ABSTRACT: ON FEBRUARY 11,1992 AT 9 AM EMPLOYEE 1 SUFFERED FATAL HEAD
INJURIES WHEN HE WAS CAUGHT BETWEEN A ROTATING ARM ON THE LUMBER STACKER
AND VERTICAL 'I' BEAM. HE WAS THE LUMBER STACKER OPERATOR AND WAS
INJURED WHILE IN THE AREA BELOW THE MACHINE. HE HAD STOPPED THE STACKER
BECAUSE OF AN INTERRUPTION IN THE FLOW OF LUMBER FROM THE SORTER. HE
AND A STICKERMAN HAD LEFT THE STACKER AND GONE TO ASSIST TWO OTHER
WORKERS. A SECOND STICKERMAN AND RELIEF STACKER OPERATOR REMAINED AT
THE STACKER CONTROLS TO OPERATE THE LUMBER TRANSFER CHAINS. WHEN THE
PROBLEM WAS CORRECTED, EMPLOYEE 1 DID NOT RETURN TO HIS OPERATING
POSITION AND THE RELIEF OPERATOR/STICKERMAN STARTED THE MACHINE. THE
STACKER COMPLETED A PARTIAL CYCLE AND STOPPED AND IN THE COURSE OF
INVESTIGATING THIS MALFUNCTION, THE CO-WORKERS DISCOVERED THAT EMPLOYEE
1 WAS CAUGHT IN THE MACHINE.

VICTIM: 001   AGE: 54   SEX: M
  DISPOSITION : FATALITY   EVENT-TYPE : CAUGHT IN OR BETWEEN
  INJ NATURE : FRACTURE   ENVIR FACTOR: PINCH POINT ACTION
  INJ SOURCE : MACHINE   HUMAN FACTOR: LOCKOUT/TAGOUT PROCED MALFUNC
  PART-OF-BODY: HEAD   HAZ SUBSTNCE: NO SUBSTANCE IMPLICATED

•••••••••••••••••••••••••••••••

Activity Number 109362046 SIC 2421 Open Date 2/19/92   NONUNION
**** ACCIDENT DATA ••••
    SUMMARY # 14330567 DATE: 2-18-92
DESCRIP:WOOD CHIPPER DESTRUCTED, BLEW APART, STRUCK OPERATOR
    ABSTRACT: EMPLOYEE HAD STARTED WOOD CHIPPER, HEARD CLICKING NOISE,
OBSERVED CHIPPER (SHORT TIME APPROXIMATELY 30 SECONDS) INTERNAL PROBLEMS
CAUSED DAMAGED TO ONE KNIFE AND PADDLES. PADDLES SHEARED THE MOUNTING
BOLTS AND BLEW UPPER HOUSING OFF OF THE MACHINE. METAL PARTS STRUCK
EMPLOYEE RESULTING IN FATAL HEAD INJURIES. EMPLOYEE (VICTIM) WAS 3' TO
5' FROM THE MORBARK, MODEL 58, CHIPPER.

VICTIM: 001   AGE: 38   SEX: M
  DISPOSITION : FATALITY   EVENT-TYPE : STRUCK BY
  INJ NATURE : BRUISE/CONTUS/ABRAS ENVIR FACTOR: FLYING OBJECT ACTION
  INJ SOURCE : MACHINE   HUMAN FACTOR: OTHER
  PART-OF-BODY: HEAD   HAZ SUBSTNCE: NO SUBSTANCE

Appendix J

Sample Tabletop Exercise

EXERCISE SCENARIO AND SEQUENCE OF EVENTS

XYZ Chemicals, Inc., produces a small line of acids for sale in the manufacturing and trade markets. XYZ stores and handles a variety of chemicals onsite for use in its own processes. In addition, XYZ products are often stored on plant grounds pending shipment to customers.

XYZ's Lake City plant is located in a neighborhood characterized by a mix of industrial and residential land uses. Local industrial facilities include two large steel plants, an oil refinery, and numerous specialty chemical plants. The XYZ facility itself is bounded on the north by a spur of the Union Railroad, on the south by the Grand River, on the west by Elm Street and a rail line, and on the east by First Street. Beyond the river on the south is an interstate highway that is heavily traveled. Within two miles north of the plant are seven schools and a hospital. Just north of the plant, across the Union Railroad tracks, is a residential neighborhood. Additional residential zones of Lake City and Middletown lie one and a half miles to the south and southwest, and three miles to the southeast.

Returning from their 12 P.M. break, two XYZ workers resume the task of transferring anhydrous hydrogen fluoride (AHF) from a pressurized rail car to a 15,000 gallon outdoor storage tank. They had allowed the transferring pump to operate unattended and find upon returning that the failure of an automatic shutoff valve resulted in a spill of approximately 1,000 gallons. The liquid AHF has begun to pool, giving rise to vapor. Inhaling these vapors, both workers suffer severe respiratory injury. Although one worker collapses immediately, the other succeeds in activating the plant safety alarm, thereby alerting the shift supervisor that an emergency has occurred at the transfer site.

The supervisor drives to investigate the accident. Smelling the strong presence of AHF vapors in the air, the supervisor stops his vehicle 200 feet from the accident site and radios the plant gate to notify the Lake City emergency response authorities by calling 911. In the act of suiting up with protective equipment, the supervisor himself collapses.

139

SEQUENCE OF EVENTS AND EXPECTED ACTIONS

EVENT 1: Plant Supervisor Calls 911

Message

From: Plant employee

To: 911

"This is a drill. There has been a chemical spill at the XYZ plant on Elm Street."

Note: No information is provided on identity of chemicals involved.

Expected Actions

- 911 makes necessary notifications, including the following:
  - ✓ Police Department
  - ✓ Fire Department
  - ✓ Emergency Medical Services
- Other notifications made, including the following:
  - ✓ State Department of Environmental Management - State Police
  - ✓ Plainville Fire Department (to activate mutual aid)
  - ✓ Middletown Hazmat Squad
  - ✓ CHEMTREC/CHEMNET
  - ✓ National Response Center

EVENT 2: Flow of Chemical Continuing At A Rapid Rate

Plant personnel evacuate. Six workers suffer eye and respiratory irritation. Condition of shift supervisor and two workers is unknown. It also not known whether all other personnel are safely out of the plant.

Expected Actions

- First Responders (whether Fire, Police, or Emergency Medical Services, EMS) assess the situation.
- Confer with plant personnel to determine identity of chemical(s).
- Count the number of evacuated personnel.

Messages

From: Plant employee

To: First responders

"The chemical leaking from tank is anhydrous hydrofluoric acid (AHF). Judging from the rate of vapor formation, it is a rapid leak."

From: Plant employee

To: First responders

"Plant personnel have evacuated. Six evacuated workers have suffered injury. Shift supervisor and two employees are known missing. Not known whether all other workers have been safely evacuated."

Expected Actions

- Police Department (upon arrival)
  - ✓ Close off access to plant.
- EMS (upon arrival)
  - ✓ Establish treatment zone in a safe area.
  - ✓ Begin examining/treating injured workers.
  - ✓ Radio for backup units.
  - ✓ Notify city hospital to expect injured.
- Fire Department (upon arrival)
  - ✓ Establish command post in a safe area.
  - ✓ Delineate "restricted areas," staging area, decontamination zone.
  - ✓ Determine personnel and equipment needs.
  - ✓ Call for additional resources as needed.
  - ✓ Squad 1 personnel (and possibly Middletown Hazmat team) suit up in protective clothing to investigate leak and injured.
  - ✓ Squad 1 approaches accident site from upwind position.
  - ✓ Spokesman issues initial press statement.

EVENT 3: Wind Observed Blowing Out Of South/Southwest at 5 MPH

Message

From: Exercise Director

To: Fire Department Incident Commander

"Winds blowing out of south/southwest at 5 MPH."

Expected Actions

- Begin consideration of evacuation option.
- Evacuation notices begin:
  - ✓ School bus company (to dispatch 3 buses)
  - ✓ Red Cross, Salvation Army
  - ✓ Lake City Civil Defense

EVENT 4: Three Additional Injured Plant Workers Discovered In Plant Powerhouse

Message

From: Exercise Director

To: Fire Department Incident Commander

"Three more injured workers have called in from plant powerhouse."

Expected Actions

- Fire Department/Middletown Hazmat personnel (with protective gear) dispatched to powerhouse to evacuate additional injured.
- Shift supervisor, two other initial injured employees evacuated by Squad 1 personnel to decontamination zone.
- Initial injured are decontaminated (as necessary).
- Initial injured are taken to EMS treatment zone.
- EMS begins triage/hospital evacuation procedures on initial injured.

EVENT 5: Because Of Valve Closure, Flow Of AHF Has Stopped; Vapor Formation Stops

Messages

None

Expected Actions

- Fire Department crew notifies Fire Department Incident Commander that leak has been stopped.
- Fire Department begins vapor suppression, pool containment procedures.
- Fire Department personnel evacuate additional injured from powerhouse to decontamination zone.
- Squad 1 members, additional injured decontaminated (as necessary).
- Additional injured taken to EMS treatment area.
- EMS begins triage/hospital evacuation procedures.

EVENT 6: Response Completed; Incident Over

Messages

From: Exercise Director

To: Fire Department Incident Commander

"The incident is over."

Expected Actions

- All response personnel notified.
- Triage/hospital evacuation completed.
- Access to plant reopened.
- Clean-up contractor(s) notified.
- Press is briefed by press spokesman, plant spokesman.

Appendix K

SAMPLE FUNCTIONAL EXERCISES

As described in the text, functional exercises consist of choosing a set of skills and practicing them to sharpen the response team. Here are some examples.

PPE

Timed donning and doffing exercises coupled with physical exercises in selected PPE ensembles. Physical exercises could include stair or ladder climbing or walks.

SPILL CONTROL

Drums

A functional exercise for drum users could be the use of spill control bands or plugs on leaking drums. A drum can be set up with a hole which is covered with a water-soluble temporary patch. When the patch erodes, the spill team is dispatched.

Pipes/Hoses

Functional exercises for specialty chemicals (chlorine, ammonia) or specialty equipment (such as feed tanks, cylinders, or special fittings) might involve the use of special response tools or special methodologies for controlling the release. These would include firefighting, chemical control, or sealant methodologies.

Booms/Pigs

Spill control functional exercises should also be performed to make employees aware of the use of various types of spill control materials and the effects of their use. Especially challenging is the use of booms or pigs to channel spills away from drains to recovery areas.

IDENTIFICATION

The incident commander may wish to set up a test which involves identifying hazardous materials from a distance or with a limited number of clues. Partial labels, incomplete shipping papers, or other clues can provide a challenge for the emergency response team.

Appendix L

SAMPLE FULL-SCALE EXERCISE

EXERCISE SCENARIO AND SEQUENCE OF EVENTS

The PQX Chemical Company plant, located on Lee Highway, manufactures a variety of corrosive, toxic, and flammable chemicals. Many of these chemicals are stored at the plant pending shipment to customers. The plant occupies 50 acres of land and is situated in an area composed of commercial, industrial, and residential buildings. The plant property is bounded on the north by Lee Highway, on the east by a rail line, on the south by Interstate 20 and on the west by the Black River. Beyond the river to the west is the Black River Estates housing development. South of Interstate 20 is the Clover Hill housing development. On the north side of Lee Highway lies a mixture of commercial and industrial buildings. East of the railroad line, there are a variety of industrial facilities. A railroad siding extends into the plant property to the outside storage area.

One clear Saturday morning, a day when the plant is not operating, a repair crew is working on replacing a section of pipe that is connected to the top of an empty tank. After disconnection, a crane is used to lower the pipe onto a flat bed truck. As the crane boom is swung over a nearby tank of liquid sulfur trioxide (SO_3), the cable snaps and drops the pipe. The falling pipe shears off the SO_3 tank's feedline between the tank wall and the first block valve. The four-inch diameter feedline leading from the tank to the process plant begins leaking immediately. An excess-flow valve between the leak and the tank limits the rate of flow to 30 gallons per minute.

Spilled liquid from the SO_3 tank collects within the containment dike surrounding the tanks. Upon contact with the moisture in the air, the spilled SO_3 vaporizes into a white mist resembling steam. The wind, coming from the northwest at 5 mph, blows the vapors directly onto the nearby repair crew that had removed the old piping. Three of the four repairmen are affected by the vapors, and two of them lose consciousness while on top of the tank they had been working on. The lone conscious repairman drags the remaining workers from the hazardous area and then runs into the main plant to report the accident.

Experiencing burning eyes and difficult breathing, he decides to remain indoors awaiting the arrival of the fire department.

SAMPLE SEQUENCE OF EVENTS

EVENT 1: Northeasterly Wind Blows The Vapor Cloud

The northeasterly wind blows the vapor cloud coming off the spilled sulfur trioxide towards the southwest across plant property. Arriving fire/rescue personnel find an unconscious person just beyond the diked area.

Simulation Message

Simulators will set off a white smoke grenade or other smoke/cloud generation device to simulate the SO_3 vapor cloud. Large portable fans may have to be employed to direct the vapors in the direction dictated by the exercise. An exercise "victim" should be lying outside of the diked area but away from the white smoke/cloud. Water from a hoseline attached to the feedline of the SO_3 tank will be flowing at a rate of 30 gpm into the diked area to simulate the leaking SO_3.

Written/Verbal Messages

None

Expected Actions

- First arriving fire/rescue units report the on-scene situation to the emergency communications center and request additional fire/rescue and police units (if necessary).

- Rescue unconscious person near diked area and provide emergency medical treatment following rescue.

- Incident commander takes command and establishes
  ✓ Command post (in safe location)
  ✓ Communications among response agencies at scene
  ✓ Staging area for in-coming apparatus
  ✓ Mechanism for on-going incident assessment

- Ensure that emergency personnel wear appropriate protective gear.

- Secure area around the incident scene.

- Attempt to locate a plant official who can identify the leaking material and provide technical expertise concerning the tank, feedline, control valves, dike, etc.

EVENT 2: Injured Repairman Reports to Fire/Rescue Personnel

The injured repairman that reported the accident advises fire/rescue personnel of the two unconscious workers on top of the tank next to the leaking tank. The repairman tells how the accident occurred and warns of the hazards of the vapors emanating from the spilled liquid.

Simulation Message

Simulators will continue to generate a white cloud to simulate the vaporizing SO_3. The simulated leak will also be continued. The repairman "victim" will be acting as if he is having trouble breathing and a burning sensation in his eyes.

Verbal Message

To: Fire/Rescue Personnel

From: Injured Repairman

"Joe and Charlie are still up on the tank. I think they've passed out. Do you see them? They were still up there when the pipe fell from the cable and hit the other tank. I don't know what that leaking stuff is, but watch out, it's nasty."

Expected Actions

- Plan strategy for the rescue of the two unconscious workers on top of the tank.

- Provide emergency medical treatment for the repairman experiencing difficult breathing and burning eyes.

- Contact CHEMTREC and/or other technical assistance organizations for assistance in identifying the leaking chemical.

- Activate the off-site emergency operations center (EOC) and notify key officials and agencies of the local government.

- Continue efforts to locate a plant official.

- Continue efforts to identify the leaking material.

- Identify strategies and options for controlling the leak.

- Arrange for specified equipment to be brought to the scene:
  - ✓ Encapsulated suits
  - ✓ Self-contained breathing apparatus
  - ✓ Environmental monitors
  - ✓ Patching/plugging materials
  - ✓ Foam
  - ✓ Diking materials
  - ✓ Emergency medical supplies

- Notify the following:
  - ✓ Community Emergency Coordinator
  - ✓ National Response Center
  - ✓ State Environmental Protection Agency

EVENT 3: The Vapor Cloud Is Approaching Interstate 20

Simulation Message

Simulators will continue to generate the white cloud but not in amounts great enough to transport the cloud to the interstate. This is to avoid obstructing the view of passing motorists not involved in the exercise. The purpose of the cloud is for realism at the actual storage tank area.

Verbal Message (via two-way radio)

To: On-Scene Incident Commander

From: Emergency Communications Center

"Motorists on Interstate 20 are reporting white smoke just north of the interstate. Could that be coming from your location?"

Expected Actions

- Initiate monitoring of vapor cloud and spill.

- Confirm vapor cloud movement.

- Close Interstate 20 downwind of the vapor cloud.

- Consider protective actions for residents south of Interstate 20.

- Request mutual aid (if necessary):
  - ✓ Fire/rescue
  - ✓ Hazardous materials team
  - ✓ Emergency medical services
  - ✓ Law enforcement

- Assist and coordinate arriving mutual aid units:
  - ✓ Brief them about incident
  - ✓ Assign tasks to them
  - ✓ Ensure they wear appropriate protective gear
  - ✓ Establish inter-organizational communications

- Establish communications between the on-scene command post and the EOC, and coordinate all response actions.

- Expand efforts to secure the area around the incident scene:
  - ✓ Roadblocks
  - ✓ Rerouting of traffic
  - ✓ Spectator control

- Establish a media center and appoint a public information officer.

EVENT 4: Location of Unconscious Workers

Fire/rescue personnel have located the two unconscious workmen on top of the tank next to the leaking tank.

Simulation Message

Simulators will continue to generate the white cloud and allow the 30 gpm flow of water into the diked area to continue. The two "victims" on top of the tank should lie still to simulate unconsciousness.

Written/Verbal Message

None

Expected Actions

- Rescue the two unconscious workers if it is decided that adequate protective gear is available at the scene for rescuers.

- Provide emergency medical treatment for the two unconscious workers following their rescue.

- Establish an on-scene triage area for injured workers and emergency response personnel.

EVENT 5: Vapor Cloud Approaches Interstate and Residential Area

The vapor cloud has moved as far as Interstate 20 and is fast approaching the Clover Hill housing development. A plant official arrives on the scene and advises the Incident Commander that the leaking product is liquid and that the 70-ton capacity tank was approximately 80 percent full prior to the accident.

Simulation Message

Simulators will continue to generate the white cloud in the area near the tanks and continue the flow of water into the diked area.

Verbal Messages

To: On-Scene Incident Commander

From: Emergency Communications Center

"Motorists are now reporting a white mist coming across the interstate from the northwest. They advise that it's irritating to their eyes and throats."

To: On-Scene Police Department Commander

From: Patrol Unit

"The vapors from your location have reached Interstate 20 and are heading towards Clover Hill. Please advise."

To: On-Scene Incident Commander

From: PQX Chemical Company Official

"The leaking product is SO_3. As of close of business yesterday, it contained approximately 55 tons of SO_3."

Expected Actions

- Evacuate Clover Hill and other nearby residences.

- Open emergency shelters for evacuees.

- Disseminate information to all emergency response personnel and agencies involved in the incident that the leaking material has been identified as liquid SO_3.

- Contact CHEMTREC and/or other technical assistance organizations to obtain the following:
  - ✓ Chemical-specific information
  - ✓ Information about associated health hazards
  - ✓ Recommended control/cleanup actions

- Ensure that protective gear is compatible with SO_3 and is worn by all emergency personnel operating in the vicinity of the leaking tank and vapor cloud.

- Continue monitoring the vapor cloud for movement and concentration.

- Identify strategies and options for reducing the quantity of vapors emanating from the spilled SO_3.

- Continue efforts to identify strategies and options for controlling the leak.

- Coordinate response efforts between the on-scene incident commander, plant officials, and the EOC.

- Provide public information concerning the following:
  - ✓ Hazards
  - ✓ Evacuation
  - ✓ Safety/precautions
  - ✓ Details of remedial actions

EVENT 6: Spilled Liquid Filling Containment Area

Despite vaporization, the diked area is filling up rapidly with spilled liquid.

Simulation Message

Simulators will continue to allow the water to flow from the hoseline into the diked area at a rate of 30 gallons per minute. The white cloud will also continue to be generated to simulate vaporization of product.

Written Message (via messenger)

To: On-Scene Incident Commander

From: Simulator

"The diked area contains a considerable amount of SO_3 and continues to fill at a rapid rate."

Expected Actions

- Arrange for the off-loading of the SO_3 from the damaged tank to other large capacity tanks that are compatible with SO_3.

- Identify strategies and options for removing the SO_3 contained within the dikes.

EVENT 7: Winds Shift; Temperature and Humidity Increase

Winds begin to shift from the northeast to the southeast. The National Weather Service's forecast calls for temperatures and humidity to increase as winds shift.

Simulation Message

Simulators will employ the use of large fans (if necessary) to simulate a wind shift so that the white cloud will blow towards the west instead of the southwest. The simulated SO_3 spill will be continued at 30 gpm.

Written Messages (via messenger)

To: On-Scene Incident Commander

From: National Weather Service

"We advise that winds will be shifting over the next 10 to 12 hours to the southeast at 3 mph. Temperatures will rise 5 to 7 degrees, and humidity will increase as well."

Verbal Messages (via two-way radio)

To: On-Scene Police Department Commander

From: Patrol Unit

"I'm at the roadblock along westbound Interstate 20. It looks as though the vapor cloud is heading more towards the west now, in the direction of Black River Estates."

To: On-Scene Incident Commander

From: Emergency Communications Center

"Citizens are reporting irritating vapors in the Black River Estates area. We've received several calls about this."

Expected Actions

- Disseminate information concerning the wind shift and weather forecast to all emergency response personnel and agencies involved in the incident.

- Evacuate the Black River Estates housing development.

- Open additional emergency shelters for evacuees.

- Expand efforts to secure the area to the west of the plant:
  - ✓ Set up roadblocks
  - ✓ Reroute traffic
  - ✓ Control spectators

EVENT 8: Unconscious Victim Discovered

An unconscious person has been spotted in a canoe floating down the Black River just west of the plant. In addition, numerous residents west of the plant have been injured.

Message

Simulators will continue their efforts to direct a white cloud towards the west. They will also continue to allow water to flow at 30 gpm into the diked area. One "victim" will lie in a slumped position in a canoe in a calm spot on the river. Several "victims" in the Black River Estates will act as though they are experiencing difficult breathing and burning eyes.

Verbal Messages (via two-way radio)

To: Emergency Medical Services Commander

From: Emergency Communications Center

"We've received a report of an unconscious person in a canoe on the Black River between the PQX plant and Interstate 20. The caller saw the canoe floating from the area affected by the vapor cloud."

To: Emergency Medical Services Commander

From: Emergency Communications Center

"Police report that in the process of evacuating Black River Estates they have found numerous persons who requested emergency medical treatment for irritated eyes and noses."

Expected Actions

- Provide emergency medical treatment for numerous injured persons.
- Rescue the unconscious canoeist.
- Evacuate the commercial/industrial area northwest of the plant.
- Expand efforts to secure the area to the northwest of the plant:
  ✓ Set up roadblocks
  ✓ Reroute traffic
  ✓ Control spectators
- Continue monitoring vapor cloud for movement and concentration.
- Apply acid-based foam (if available) to the surface of the contained SO_3 to prevent the release of hazardous vapors.

EVENT 9: Tank cars Brought in To Begin Off-Loading Damaged Tank

Two railroad tank cars have been brought onto the siding next to the SO_3 tank. Off-loading operations will be difficult due to the presence of spilled liquid within the diked area around the tank.

Simulation Message

Simulators will continue to allow water to flow into the diked area and will continue to generate a white cloud and direct it towards the west.

Verbal Messages

To: Plant Official

From: Railroad Engineer

"How should I position the two empty tank cars for off-loading operations?"

To: On-Scene Incident Commander

From: Senior Fire Department Officer

"We're going to have a difficult time gaining access to the unloading outlet on the tank with all this liquid SO_3 around the base of the tank. It would be unsafe to have anyone walk through the liquid, even if they're wearing Level A protective gear."

Expected Actions

- Identify strategies and options for gaining access to the unloading outlet on the SO_3 tank without endangering the lives of the personnel assigned the task.

- Off-load the SO_3 from the tank to the railroad tank cars.

- Continue monitoring the vapor cloud for movement and concentration.

EVENT 10: Off-Loading Completed; Vapors Continue to Emanate from Spill

The product has been completely off-loaded from the tank to the railcars, thus ending the leak. Vapors, however, continue to be given off from the spilled liquid within the dikes.

Simulation Message

Simulators will continue to generate a white cloud and direct it towards the west until actions are taken to prevent the vaporization of product and/or the product is pumped to tanks.

Verbal Messages

To: On-Scene Incident Commander

From: Senior Fire Department Officer

"We just finished off-loading the SO_3 to the tankcars. The leak has stopped."

To: On-Scene Incident Commander

From: Senior Fire Department Officer

"The contained liquid is still vaporizing."

Expected Actions

- Inform all emergency responders and agencies involved in the incident that the leak has been stopped but that hazardous vapors continue to be generated from the spilled material.

- Continue monitoring the vapor cloud for movement and concentration.

- Apply acid-based foam (if available) to the surface of the contained SO_3 to prevent the release of hazardous vapors (if not already done).

- Pump liquid SO_3 from the containment dikes into rail tank cars or other compatible tanks.

- Following the complete elimination of hazards, consider post-incident operations including:
  - ✓ Cleanup
  - ✓ Decontamination of personnel, equipment, apparatus, and property
  - ✓ Removal and disposal of hazardous wastes
  - ✓ Re-entry of evacuees to residential areas
  - ✓ Opening of roads and the evacuated commercial/industrial area
  - ✓ Continued air monitoring

Appendix M

NIOSH/OSHA/USCG/EPA
OCCUPATIONAL SAFETY AND HEALTH
GUIDANCE MANUAL FOR HAZARDOUS
WASTE SITE ACTIVITIES (1985)
(APPENDIX D)

Appendix D. Sample Decontamination Procedures for Three Typical Levels of Protection[a]

F.S.O.P. No. 7

Process: <u>DECONTAMINATION PROCEDURES</u>

INTRODUCTION

1.1 The objective of these procedures is to minimize the risk of exposure to hazardous substances. These procedures were derived from the U.S. Environmental Protection Agency, Office of Emergency and Remedial Response's (OERR), "Interim Standard Operating Safety Guides (revised Sep. 82)". This version of the guides is in a format that is more appropriate for use in the field.

1.2 Protective equipment must be worn by personnel when response activities involve known or suspected hazardous substances. The procedures for decontaminating personnel upon leaving the contaminated area are addressed for each of the EPA, OERR designated levels of protection. The procedures given are for the maximum and minimum amount of decontamination used for each level of protection.

1.3 The maximum decontamination procedures for all levels of protection consist of specific activities at nineteen stations. Each station emphasizes an important aspect of decontamination. When establishing a decontamination line, each aspect should be incorporated separately or combined with other aspects into a procedure with fewer steps (such as the Minimum Decontamination Procedures).

1.4 Decontamination lines are site specific since they are dependent upon the types of contamination and the type of work activities on site. A cooling station is sometimes necessary within the decontamination line during hot weather. It is usually a location in a shaded area in which the wind can help to cool personnel. In addition, site conditions may permit the use of cooling devices such as cool water hose, ice packs, cool towels, etc. When the decontamination line is no longer required, contaminated wash and rinse solutions and contaminated articles must be contained and disposed of as hazardous wastes in compliance with state and federal regulations.

---

[a] **Source:** Excerpted from *Field Standard Operating Procedures for the Decontamination of Response Personnel (FSOP 7)*. EPA Office of Emergency and Remedial Response, Hazardous Response Support Division, Washington, DC. January 1985.

P.S.O.P. No. 7

PROCESS   DECON PROCEDURES

MAXIMUM DECONTAMINATION LAYOUT

LEVEL A PROTECTION

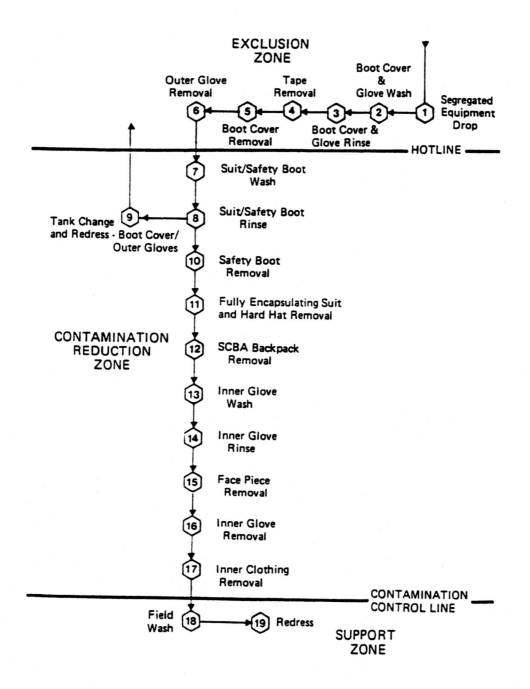

F.S.O.P. No. 7

PROCESS   DECON PROCEDURES

MAXIMUM DECONTAMINATION LAYOUT

LEVEL B PROTECTION

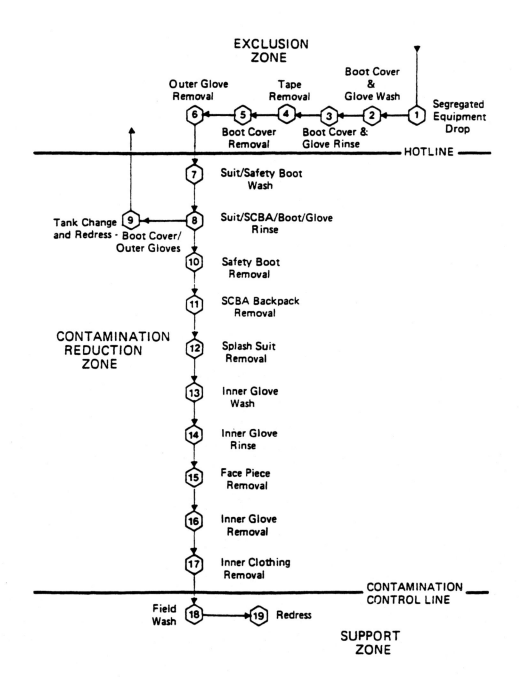

P.S.O.P. No. 7

PROCESS   DECON PROCEDURES

MAXIMUM DECONTAMINATION LAYOUT

LEVEL C PROTECTION

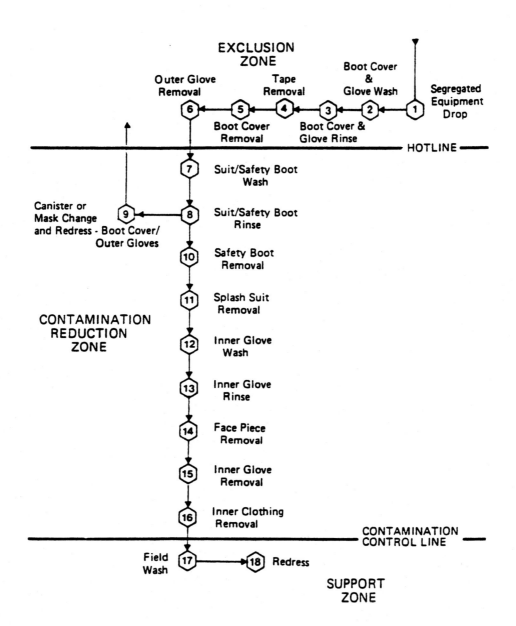

F.S.O.P. No. 7

PROCESS   DECON PROCEDURES

MINIMUM DECONTAMINATION LAYOUT

LEVELS A & B PROTECTION

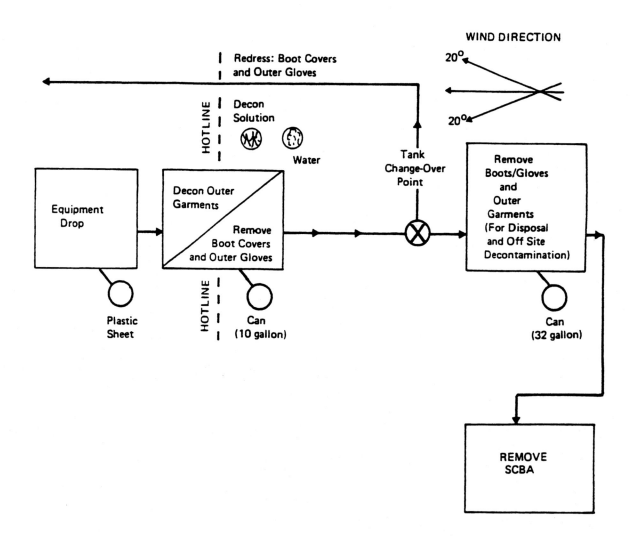

F.S.O.P. No. 7

PROCESS   DECON PROCEDURES

MINIMUM DECONTAMINATION LAYOUT

LEVEL C PROTECTION

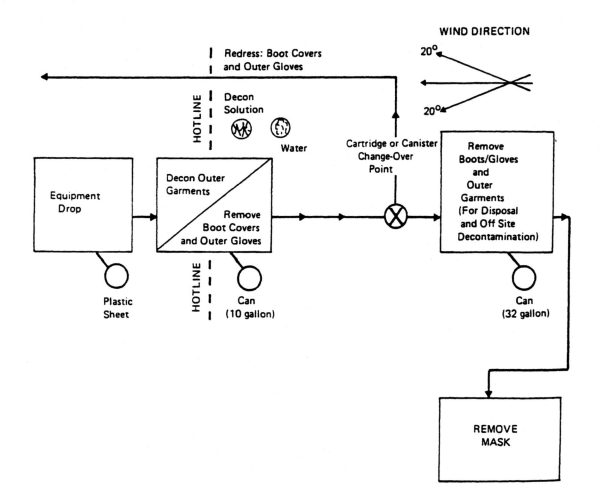

EQUIPMENT NEEDED TO PERFORM MAXIMUM DECONTAMINATION MEASURES FOR LEVELS A, B, AND C

Station 1:   a. Various Size Containers
           b. Plastic Liners
           c. Plastic Drop Cloths

Station 2:   a. Containers (20-30 Gallons)
           b. Decon Solution or Detergent Water
           c. 2-3 Long-Handled, Soft-Bristled
              Scrub Brushes

Station 3:   a. Containers (20-30 Gallons)
                OR
              High-Pressure Spray Unit
           b. Water
           c. 2-3 Long-Handled, Soft-Bristled
              Scrub Brushes

Station 4:   a. Containers (20-30 Gallons)
           b. Plastic Liners

Station 5:   a. Containers (20-30 Gallons)
           b. Plastic Liners
           c. Bench or Stools

Station 6:   a. Containers (20-30 Gallons)
           b. Plastic Liners

Station 7:   a. Containers (20-30 Gallons)
           b. Decon Solution or Detergent Water
           c. 2-3 Long-Handled, Soft-Bristled
              Scrub Brushes

Station 8:   a. Containers (20-30 Gallons)
                OR
              High-Pressure Spray Unit
           b. Water
           c. 2-3 Long-Handled, Soft-Bristled
              Scrub Brushes

Station 9:   a. Air Tanks or Face Masks and
              Cartridge Depending on Level
           b. Tape
           c. Boot Covers
           d. Gloves

Station 10: a. Containers (20-30 Gallons)
           b. Plastic Liners
           c. Bench or Stools
           d. Boot Jack

Station 11: a. Rack
           b. Drop Cloths
           c. Bench or Stools

Station 12: a. Table

Station 13: a. Basin or Bucket
           b. Decon Solution
           c. Small Table

Station 14: a. Water
           b. Basin or Bucket
           c. Small Table

Station 15: a. Containers (20-30 Gallons)
           b. Plastic Liners

Station 16: a. Containers (20-30 Gallons)
           b. Plastic Liners

Station 17: a. Containers (20-30 Gallons)
           b. Plastic Liners

Station 18: a. Water
           b. Soap
           c. Small Table
           d. Basin or Bucket
           e. Field Showers
           f. Towels

Station 19: a. Dressing Trailer is Needed in
              Inclement Weather
           b. Tables
           c. Chairs
           d. Lockers
           e. Cloths

EQUIPMENT NEEDED TO PERFORM MINIMUM DECONTAMINATION MEASURES FOR LEVELS A, B, AND C

Station 1:   a. Various Size Containers
           b. Plastic Liners
           c. Plastic Drop Cloths

Station 2:   a. Containers (20-30 Gallons)
           b. Decon Solution
           c. Rinse Water
           d. 2-3 Long-Handled, Soft-Bristled
              Scrub Brushes

Station 3:   a. Containers (20-30 Gallons)
           b. Plastic Liners
           c. Bench or Stools

Station 4:   a. Air Tanks or Masks and
               Cartridges Depending Upon Level
           b. Tape
           c. Boot Covers
           d. Gloves

Station 5:   a. Containers (20-30 Gallons)
           b. Plastic Liners
           c. Bench or Stools

Station 6:   a. Plastic Sheets
           b. Basin or Bucket
           c. Soap and Towels
           d. Bench or Stools

Station 7:   a. Water
           b. Soap
           c. Tables

FSOP 7: MAXIMUM MEASURES FOR LEVEL A DECONTAMINATION

Station 1: Segregated Equipment Drop

1. Deposit equipment used on site (tools, sampling devices and containers, monitoring instruments, radios, clipboards, etc.) on plastic drop cloths or in different containers with plastic liners. During hot weather operations, a cool down station may be set up within this area.

Station 2: Boot Cover and Glove Wash

2. Scrub outer boot covers and gloves with decon solution or detergent/water.

Station 3: Boot Cover and Glove Rinse

3. Rinse off decon solution from station 2 using copious amounts of water.

Station 4: Tape Removal

4. Remove tape around boots and gloves and deposit in container with plastic liner.

Station 5: Boot Cover Removal

5. Remove boot covers and deposit in container with plastic liner.

Station 6: Outer Glove Removal

6. Remove outer gloves and deposit in container with plastic liner.

Station 7: Suit and Boot Wash

7. Wash encapsulating suit and boots using scrub brush and decon solution or detergent/water. Repeat as many times as necessary.

Station 8: Suit and Boot

8. Rinse off decon solution using water. Repeat as many times as necessary.

Station 9: Tank Change

9. If an air tank change is desired, this is the last step in the decontamination procedure. Air tank is exchanged, new outer gloves and boot covers donned, and joints taped. Worker returns to duty.

Station 10: Safety Boot Removal

10. Remove safety boots and deposit in container with plastic liner.

Station 11: Fully Encapsulating Suit and Hard Hat Removal

11. Fully encapsulated suit is removed with assistance of a helper and laid out on a drop cloth or hung up. Hard hat is removed. Hot weather rest station maybe set up within this area for personnel returning to site.

Station 12: SCBA Backpack Removal

12. While still wearing facepiece, remove backpack and place on table. Disconnect hose from regulator valve and proceed to next station.

Station 13: Inner Glove Wash

13. Wash with decon solution that will not harm the skin. Repeat as often as necessary.

Station 14: Inner Glove Rinse

14. Rinse with water. Repeat as many times as necessary.

Station 15: Face Piece Removal

15. Remove face piece. Deposit in container with plastic liner. Avoid touching face with fingers.

Station 16: Inner Glove Removal

16. Remove inner gloves and deposit in container with liner.

FSOP 7: MAXIMUM MEASURES FOR LEVEL A DECONTAMINATION

Station 17: Inner Clothing Removal

17. Remove clothing and place in lined container. Do not wear inner clothing off-site since there is a possibility that small amounts of contaminants might have been transferred in removing the fully-encapsulating suit.

Station 18: Field Wash

18. Shower if highly toxic, skin-corrosive or skin-absorbable materials are known or suspected to be present. Wash hands and face if shower is not available.

Station 19: Redress

19. Put on clean clothes.

FSOP 7: MINIMUM MEASURES FOR LEVEL A DECONTAMINATION

Station 1: Equipment Drop

1. Deposit equipment used on-site (tools, sampling devices and containers, monitoring instruments, radios, clipboards, etc.) on plastic drop cloths. Segregation at the drop reduces the probability of cross contamination. During hot weather operations, cool down stations maybe set up within this area.

Station 2: Outer Garment, Boots, and Gloves Wash and Rinse

2. Scrub outer boots, outer gloves and fully-encapsulating suit with decon solution or detergent and water. Rinse off using copious amounts of water.

Station 3: Outer Boot and Glove Removal

3. Remove outer boots and gloves. Deposit in container with plastic liner.

Station 4: Tank Change

4. If worker leaves Exclusion Zone to change air tank, this is the last step in the decontamination procedure. Worker's air tank is exchanged, new outer gloves and boot covers donned, joints taped, and worker returns to duty.

Station 5: Boot, Gloves and Outer Garment Removal

5. Boots, fully-encapsulating suit, inner gloves removed and deposited in separate containers lined with plastic.

Station 6: SCBA Removal

6. SCBA backpack and facepiece is removed (avoid touching face with fingers). SCBA deposited on plastic sheets.

Station 7: Field Wash

7. Hands and face are thoroughly washed. Shower as soon as possible.

FSOP 7: MAXIMUM MEASURES FOR LEVEL B DECONTAMINATION

Station 1: Segregated Equipment Drop

1. Deposit equipment used on site (tools, sampling devices and containers, monitoring instruments, radios, clipboards, etc.) on plastic drop cloths or in different containers with plastic liners. Segregation at the drop reduces the probability of cross-contamination. During hot weather operations, cooldown stations may be set up within this area.

Station 2: Boot Cover and Glove Wash

2. Scrub outer boot covers and gloves with decon solution or detergent and water.

Station 3: Boot Cover and Glove Rinse

3. Rinse off decon solution from station 2 using copious amounts of water.

Station 4: Tape Removal

4. Remove tape around boots and gloves and deposit in container with plastic liner.

Station 5: Boot Cover Removal

5. Remove boot covers and deposit in container with plastic liner.

Station 6: Outer Glove removal

6. Remove outer gloves and deposit in container with plastic liner.

Station 7: Suit and Safety Boot Wash

7. Wash chemical-resistant splash suit, SCBA, gloves and safety boots. Scrub with long-handle scrub brush and decon solution. Wrap SCBA regulator (if belt mounted type) with plastic to keep out water. Wash backpack assembly with sponges or cloths.

Station 8: Suit, SCBA, Boot, and Glove Rinse

8. Rinse off decon solution using copious amounts of water.

Station 9: Tank Change

9. If worker leaves exclusion zone to change air tank, this is the last step in the decontamination procedure. Worker's air tank is exchanged, new outer gloves and boot covers donned, and joints taped. Worker returns to duty.

Station 10: Safety Boot Removal

10. Remove safety boots and deposit in container with plastic liner.

Station 11: SCBA Backpack Removal

11. While still wearing facepiece, remove backpack and place on table. Disconnect hose from regulator valve.

Station 12: Splash Suit Removal

12. With assistance of helper, remove splash suit. Deposit in container with plastic liner.

Station 13: Inner Glove Wash

13. Wash inner gloves with decon solution.

Station 14: Inner Glove Rinse

14. Rinse inner gloves with water.

Station 15: Face Piece Removal

15. Remove face piece. Deposit in container with plastic liner. Avoid touching face with fingers.

Station 16: Inner Glove Removal

16. Remove inner gloves and deposit in container with liner.

FSOP 7: MAXIMUM MEASURES FOR LEVEL B DECONTAMINATION

Station 17: Inner Clothing Removal	17. Remove inner clothing. Place in container with liner. Do not wear inner clothing off-site since there is a possibility that small amounts of contaminants might have been transferred in removing the fully-encapsulating suit.
Station 18: Field Wash	18. Shower if highly toxic, skin-corrosive or skin-absorbable materials are known or suspected to be present. Wash hands and face if shower is not available.
Station 19: Redress	19. Put on clean clothes.

FSOP 7: MINIMUM MEASURES FOR LEVEL B DECONTAMINATION

Station 1: Equipment Drop	1. Deposit equipment used on-site (tools, sampling devices and containers, monitoring instruments, radios, clipboards, etc.) on plastic drop cloths. Segregation at the drop reduces the probability of cross contamination. During hot weather operations, cool down station may be set up within this area.
Station 2: Outer Garment, Boots, and Gloves Wash and Rinse	2. Scrub outer boots, outer gloves and chemical-resistant splash suit with decon solution or detergent water. Rinse off using copious amounts of water.
Station 3: Outer Boot and Glove Removal	3. Remove outer boots and gloves. Deposit in container with plastic liner.
Station 4: Tank Change	4. If worker leaves exclusive zone to change air tank, this is the last step in the decontamination procedure. Worker's air tank is exchanged, new outer gloves and boot covers donned, joints taped, and worker returns to duty.
Station 5: Boot, Gloves and Outer Garment Removal	5. Boots, chemical-resistant splash suit, inner gloves removed and deposited in separate containers lined with plastic.
Station 6: SCBA Removal	6. SCBA backpack and facepiece is removed. Avoid touching face with finger. SCBA deposited on plastic sheets.
Station 7: Field Wash	7. Hands and face are thoroughly washed. Shower as soon as possible.

FSOP 7: MAXIMUM MEASURES FOR LEVEL C DECONTAMINATION

Station 1: Segrated Equipment Drop

1. Deposit equipment used on site (tools, sampling devices and containers, monitoring instruments, radios, clipboards, etc.) on plastic drop cloths or in different containers with plastic liners. Segregation at the drop reduces the probability of cross contamination. During hot weather operations, a cool down station may be set up within this area.

Station 2: Boot Cover and Glove Wash

2. Scrub outer boot covers and gloves with decon solution or detergent and water.

Station 3: Boot Cover and Glove Rinse

3. Rinse off decon solution from station 2 using copious amounts of water.

Station 4: Tape Removal

4. Remove tape around boots and gloves and deposit in container with plastic liner.

Station 5: Boot Cover Removal

5. Remove boot covers and deposit in containers with plastic liner.

Station 6: Outer Glove Removal

6. Remove outer gloves and deposit in container with plastic liner.

Station 7: Suit and Boot Wash

7. Wash splash suit, gloves, and safety boots. Scrub with long-handle scrub brush and decon solution.

Station 8: Suit and Boot, and Glove Rinse

8. Rinse off decon solution using water. Repeat as many times as necessary.

Station 9: Canister or Mask Change

9. If worker leaves exclusion zone to change canister (or mask), this is the last step in the decontamination procedure. Worker's canister is exchanged, new outer gloves and boot covers donned, and joints taped worker returns to duty.

Station 10: Safety Boot Removal

10. Remove safety boots and deposit in container with plastic liner.

Station 11: Splash Suit Removal

11. With assistance of helper, remove splash suit. Deposit in container with plastic liner.

Station 12: Inner Glove Rinse

12. Wash inner gloves with decon solution.

Station 13: Inner Glove Wash

13. Rinse inner gloves with water.

Station 14: Face Piece Removal

14. Remove face piece. Deposit in container with plastic liner. Avoid touching face with fingers.

Station 15: Inner Glove Removal

15. Remove inner gloves and deposit in lined container.

FSOP 7:   MAXIMUM MEASURES FOR LEVEL C DECONTAMINATION

Station 16:	Inner Clothing Removal	16.	Remove clothing soaked with perspiration and place in lined container. Do not wear inner clothing off-site since there is a possibility that small amounts of contaminants might have been transferred in removing the fully-encapsulating suit.
Station 17:	Field Wash	17.	Shower if highly toxic, skin-corrosive or skin-absorbable materials are known or suspected to be present. Wash hands and face if shower is not available.
Station 18:	Redress	18.	Put on clean clothes.

FSOP 7:   MINIMUM MEASURES FOR LEVEL C DECONTAMINATION

Station 1:	Equipment Drop	1.	Deposit equipment used on-site (tools, sampling devices and containers, monitoring instruments, radios, clipboards, etc.) on plastic drop cloths. Segregation at the drop reduces the probability of cross contamination. During hot weather operations, a cool down station may be set up within this area.
Station 2:	Outer Garment, Boots, and Gloves Wash and Rinse	2.	Scrub outer boots, outer gloves and splash suit with decon solution or detergent water. Rinse off using copious amounts of water.
Station 3:	Outer Boot and Glove Removal	3.	Remove outer boots and gloves. Deposit in container with plastic liner.
Station 4:	Canister or Mask Change	4.	If worker leaves exclusive zone to change canister (or mask), this is the last step in the decontamination procedure. Worker's canister is exchanged, new outer gloves and boot covers donned, joints taped, and worker returns to duty.
Station 5:	Boot, Gloves and Outer Garment Removal	5.	Boots, chemical-resistant splash suit, inner gloves removed and deposited in separate containers lined with plastic.
Station 6:	Face Piece Removal	6.	Facepiece is removed. Avoid touching face with fingers, Facepiece deposited on plastic sheet.
Station 7:	Field Wash	7.	Hands and face are thoroughly washed. Shower as soon as possible.

Appendix N

GUIDELINES FOR DECONTAMINATION OF FIRE FIGHTERS AND THEIR EQUIPMENT FOLLOWING HAZARDOUS MATERIALS INCIDENTS

Prepared by the Dangerous Goods Subcommittee of the Canadian Association of Fire Chiefs[1]

NOTE TO READERS

The contents of this booklet are based on information and advice believed to be accurate and reliable. The Canadian Association of Fire Chiefs, Inc., its officers and members jointly and severally, make no guarantee and assume no liability in connection with this booklet. Moreover, it should not be assumed that every acceptable procedure is included or that special circumstances may not warrant modified or additional procedures.

The user should be aware that changing technology or regulations may require a change in the recommended procedures contained herein. Appropriate steps should be taken to ensure that the information is current when used.

The suggested procedures should not be confused with any federal, provincial, state, municipal, or insurance requirements, or with national safety codes.

---

[1]Used by permission of the Canadian Association of Fire Chiefs, Ottawa, Ontario, Canada.

INTRODUCTION

The number of hazardous materials incidents to which the fire service is called increases year by year. At each of these incidents, there is a good chance that the responding fire fighters may become contaminated with the hazardous material. Frequently, however, the matter of decontamination is never thought of, or is only performed cursorily.

The Dangerous Goods Subcommittee of the Canadian Association of Fire Chiefs has therefore undertaken to prepare a series of guidelines for decontamination, to be adopted by any fire department that wishes to do so, either as printed here or with local variations due to their own circumstances. Note that these are indeed *guidelines, not standards.*

The procedures listed are designed so they can be carried out by any department, rural or urban, volunteer or full-time, with a minimum of investment in special equipment.

The I.A.F.C. Hazardous Materials Committee (among many others) has reviewed this booklet and endorses the philosophies contained in it. However, they share the disclaimer caution on page 155.

BACKGROUND OF THE STUDY AND RATIONALE FOR ITS CONCLUSIONS

In 1986 the Dangerous Goods Subcommittee carried out an extensive study of fire services across the world to investigate the different approaches to decontamination of fire fighters following hazardous materials incidents.

Procedures were reviewed in detail from North America (Phoenix, San Francisco, Colorado, Metropolitan Toronto area) and England (Hampshire, Cambridgeshire, Greater Manchester, London, and the Home Office Guidelines). In addition, information was requested from Hong Kong, New Zealand, Australia, the People's Republic of China, France, Germany, Switzerland, Italy, Sweden, and the Netherlands. From these latter countries no formal replies were received, however, and indications are that decontamination procedures are either limited or absent. Japanese fire officials wrote back to indicate that they were studying various options but had not yet finalized any procedures.

Many magazine articles from the various periodicals published for the European and North American fire service market were reviewed. Furthermore, members consulted with chemical manufacturers, nuclear medicine physicists, hazardous waste disposal companies, industrial hygienists, toxicologists, and various jurisdictional agencies such as the Atomic Energy Control Board as well as the provincial and federal Ministries of Health, Labour, and Environment.

All this research let to the conclusion that there were three basic philosophies in existence:

1. Wet and Dry Procedures

2. Dispose or Retain Runoff

3. Severity and Type of the Material

The idea of a dry procedure made sense because of easier containment and no reaction with water. A single wet procedure, however, was not deemed to be sufficiently comprehensive; on the other hand, some wet procedures called for making up solutions of a variety of chemicals, and these were deemed to be too complicated for use by every fire department.

The dispose-or-retain philosophy is usually considered a wet procedure; however, to base one's procedures solely on the concern about runoff appeared to be overly simplistic. Note, however, that concerns about runoff are addressed later in this document.

The third philosophy was examined in more detail. The methods of determining the severity and type of material were defined by various fire departments along the following lines:

1. By UN class

2. By effects on the environment

3. By chemical characteristics

4. By physiological effects on people

5. By broad groups

The first alternative was deemed unsuitable because there are too many variables within a class (e.g., some flammable liquids are highly toxic, others are not). The second alternative was deemed to be of secondary importance to fire fighter safety (see the section on Environmental Considerations later in this document). It was found that alternatives 3 and 4 could in fact be used to arrive at alternative 5, and this was the route taken to define the procedures in this document.

The procedures that were developed are therefore broken down as follows:

*   Three general procedures for light, medium, and extreme hazards

*   Two specific procedures for substances that do not fit into the three general groups above (although they share many common factors)

*   One initial routine performed in some cases prior to the start of one of the other procedures

The Sub-Committee realizes that if the procedures listed here are to be completed thoroughly, a number of decontamination operatives and a Decontamination Officer to oversee them are needed. Typically, this will require the services of at least one fire fighting company. Attention to detail and careful execution of all steps should lead to successful and safe completion of fire fighter decontamination.

DECONTAMINATION: GENERAL OBSERVATIONS

Six levels of decontamination are outlined. The incident commander will determine which level is applicable for the substance involved, using any reference sources that may state the applicable level. In the absence of such sources, advice should be sought from experts such as toxicologists, chemical company representatives, CANUTEC, CHEMTREC, etc.

The levels are listed below:

A   light hazards

B   medium hazards

C   extreme hazards

D   dry contamination for water-reactive and certain dry substances

E   etiologic agents and certain dry pesticides and poisons

R   radioactive materials

Note that A-level decontamination, the most common, need only be done at the station. However, other levels need to be started at the incident scene and should be continued on return to the station.

C-level decontamination, the most stringent level for the most toxic substances, may involve the destruction of all clothes worn.

In a few cases, scrubbing of clothes must be done while wearing SCBA as vapors released during cleaning may be harmful.

D-level decontamination is almost always followed by one of the other levels of decontamination, which will be dependent upon the substance involved.

The procedures should be initiated if personnel are known or suspected to have been directly exposed to the chemical or its vapors, products of combustion, etc.

Officers should be aware of any cuts, wounds, lesions, or abrasions that their crews may have. If the apparatus is sent to an incident involving hazardous materials, such personnel should wherever possible exercise special care to avoid the chance of contamination through such wounds. Chemicals absorbed through the skin will be absorbed much faster if the skin is cut or abraded, thereby presenting a serious health hazard.

Adequate awareness is necessary to realize when decontamination will be required, so that early action can be taken to bring to the scene the equipment and manpower resources needed to set up and staff the decontamination area.

DECONTAMINATION PROCEDURE
LEVEL A: LIGHT HAZARDS

On Return to Station

1. Wash down all fire fighting clothes with a mild (1 to 2 percent) trisodium phosphate solution. Rinse with water.

2. Wash down SCBA cylinders and harnesses with a mild trisodium phosphate solution. Take care to wipe, not scrub, around regulator assembly. Rinse with clean water. If damage is suspected to any part of the unit, make sure it is sent for service.

3. Scrub hands and face with soap and water.

4. Note: Where the scrubbing of the fire clothes may release harmful vapors caught in the fibers, it may be necessary to wear breathing apparatus while washing down fire clothes. In these cases,

monitor the atmosphere around the washing area. Release of vapors may indicate commercial cleaning is required.

DECONTAMINATION PROCEDURE
LEVEL B: MEDIUM HAZARDS

At the Scene

1. Do not remove SCBA facepiece. Place helmet on back of neck.

2. Assistant must flush fire fighter downwards from head to toe with copious amounts of low pressure water from open end of firehose. Include inside and outside of helmet, mask, harness, boots down from the top, and inside of coat-wrists to the cuff.

3. Do not smoke, eat, drink, or touch face.

On Return to Station

4. Place apparatus temporarily out of service.

5. Remove all fire fighting clothes (coat, belt, boots, helmet, etc.). If possible, remove liner from helmet. Scrub all items, including the helmet liner, inside and out, with a mild (1 to 2 percent) trisodium phosphate solution. Then flush copiously with water.

6. Note: Where the scrubbing of the fire clothes might release harmful vapors caught in the fibers, it may be necessary to wear breathing apparatus while washing down fire clothes. In these cases, monitor the atmosphere around the washing area. Release of vapors may indicate commercial cleaning is required.

7. Scrub all other protective gear such as gloves and breathing apparatus items likewise. Be sure to flush out gloves with water. If SCBA is stored in its case while returning from incident, scrub the case also.

8. Remove all clothing worn at the scene, including underwear, and place in garbage bag for laundering and/or dry cleaning (preferably the latter). Take all garbage bags with contaminated clothing to a place where they can be cleaned separately from other garments.

9. Shower, scrubbing all of the body with soap and water, with particular emphasis on areas around the mouth and nostrils and under fingernails. Shampoo hair and thoroughly clean mustache if you have one.

10. Do not smoke, drink, eat, touch face, or void until step 8 is completed.

11. Put on clean clothes.

12. Do not put apparatus back in service until cleanup is completed.

To Change SCBA Cylinders at the Scene

Flush empty cylinder and surrounding area of the fire fighter's back with copious amounts of low pressure water from open end of firehose. Also flush facepiece and breathing tube to prevent inhalation of harmful materials when regulator is disconnected.

Wear gauntlet-type gloves, such as those used by linemen, when changing cylinders. Flush empty cylinder and surrounding area of fire fighter's back with copious amounts of low pressure water from open end of firehose. Also flush facepiece and breathing tube to prevent inhalation of harmful materials when regulator is disconnected. Flush gloves after use and before removing them.

DECONTAMINATION PROCEDURE
LEVEL C: EXTREME HAZARDS

At the Scene

1. Do not remove SCBA facepiece. Place helmet on back of neck.

2. Assistant, wearing turnout gear and SCBA (plus disposable chemical suit wherever possible), must flush fire fighter downwards from head to toe with copious amounts of low pressure water from open end of firehose. Include inside and outside of helmet, mask, harness, boots down from the top, and inside of coat-wrists to the cuff.

3. Do not smoke, eat, drink, or touch face.

4. Put SCBA, used cylinders, and any equipment (including hoses and tarps) suspected or known to be contaminated in garbage bags. Seal bags and return them to the station. Where circumstances permit, remove and bag firegear also.

On Return to Station

5. Put bags returned from incident scene in exterior cordoned-off area away from public access. Place apparatus out of service.

6. Strip completely. Place all clothing (firegear and personal clothing) in plastic garbage bags. Place portable radios in a separate bag. Seal bags and place in exterior cordoned-off area.

7. Arrange for the supply of a number of steel drums. Upon their arrival, seal garbage bags with contaminated items into drums. Mark drums and place in exterior cordoned-off area, minimum 5-meter radius.

8. Arrange for the drums to be picked up and the contents analyzed. Some or all items may be destroyed, some may be able to be decontaminated and returned.

9. Shower, scrubbing all of the body with soap and water, with particular emphasis on areas around the mouth and nostrils and under fingernails. Shampoo hair. Thoroughly clean mustache if you have one.

Special Attention for Radioactive Incidents:

After showering, scan entire body with a radiation contamination monitor, paying special attention to hair, hands, and fingernails. Hold monitor approximately 3 cm from body. If any reading beyond normal background level is detected, the fire fighter should shower again, scrubbing with more soap than before.

10. Do not smoke, drink, eat, touch face, or void until step 9 is completed.

11. Put on clean clothes.

12. Report to hospital for medical examination. Inform physician which hazardous material was involved.

To Change SCBA Cylinders at the Scene

Flush empty cylinder and surrounding area of fire fighter's back with copious amounts of low pressure water from open end of firehose. Also flush facepiece and breathing tube to prevent inhalation of harmful material when regulator is disconnected.

Wear gauntlet-type rubber gloves, such as those used by linemen, when changing cylinders. Flush gloves after use before removing them.

Place empty cylinder in black plastic garbage bag and seal for subsequent decontamination.

The person doing the flushing and cylinder-changing must wear turnout gear and SCBA, plus a disposable chemical suit if available.

Special Note

Where circumstances, local climate, and available resources permit, the performance of *all* steps at the scene (instead of performing steps 5 to 11 at the station) is preferable. The procedure is outlined as shown, however, in recognition of the fact that for many departments this will usually be impossible to achieve.

DECONTAMINATION PROCEDURE
LEVEL D: WATER-REACTIVE HAZARDS

At the Scene

1. Set up a suitable vacuum cleaner with power supply. Provide a dry brush and a containment capture method for materials falling off the contaminated personnel. Assistants must don full turnout gear and SCBA, plus disposable chemical suits if available and appropriate.

2. **If this is a radiation incident:**

   The fire fighters suspected of being contaminated will be scanned carefully with a radiation monitor suitable for detecting surface contamination. All parts of their clothing and personal equipment will be scanned, including the soles of the boots. If no readings are found, the personnel that have been checked can leave the decontamination area.

3. If not a radiation incident, or if the fire fighter was not found to be radioactively contaminated:

    Stand fire fighter in center of contaminant area, clean helmet, and place on back of neck. Then clean inside of helmet.

4. Commence cleaning from head downwards. Include all external areas. Slacken SCBA harness to allow cleaning behind straps and backplate. Likewise, loosen the hose-key belt and clean behind it.

5. When the fire fighter has been fully vacuumed or brushed off, he will step out of the containment area. As he does so, his boots, including the soles, must be cleaned off so any contaminant will remain within the containment area.

6. Procedures will then continue as follows:

    *   Radioactive incident: go to Level R routine

    *   Etiological or dry pesticide incident: go to Level E routine

    *   Other incidents: go to Level B routine (unless advice is received that Level C is more appropriate)

7. All used filters and collected waste are to be placed in a garbage bag, sealed and tagged, and disposed of in a manner acceptable to the agency having jurisdiction.

DECONTAMINATION PROCEDURE
LEVEL E: ETIOLOGIC HAZARDS

Special Equipment Required

A presentation spray can (such as used for pesticide spraying), bleach concentrate, orange garbage bags, black garbage bags, sterilization bags as used by hospital laundries, and a box of surgical masks.

At the Scene

1. Make up a 5 percent to 6 percent bleach solution in the spray can. Take note of the bleach concentrate percentage when calculating the makeup of the solution. Many brands as purchased in the store are already 6 percent.

2. Flush the fire fighter downwards from head to toe with low pressure water from a firehose. SCBA facepiece can now be removed. Place helmets in black plastic garbage bag(s) and seal. Place surgical mask on the fire fighter.

3. Spray the fire fighter's boots (but not bunker gear) and any tools, hoses, and other equipment used (except for portable radios) with the bleach solution in the spray can. Leave for 10 minutes, then flush with water.

4. Remove SCBA. Place in black plastic garbage bag and seal. Remove fire fighter's fire coat and gloves. Place in orange plastic garbage bag and seal. Remove any portable radio worn. Place in black plastic garbage bag and seal. Discard surgical masks.

5. Do not smoke, eat, drink, or touch face.

6. Before leaving the scene, a fire fighter wearing SCBA should attempt to spray as much of the ground exposed to the material and the wash-down water as possible with the bleach solution. Then flush the outside of the spray can with clean water.

7. Before leaving the scene, seal the orange garbage bags into the sterilization bags.

On Return to Station

8. Place apparatus temporarily out of service.

9. One fire fighter should dress in firegear and SCBA, and in an outside area perform the following tasks:

   - Open the black plastic garbage bags, wipe all helmets, portable radios, SCBA sets, and used cylinders with a rag lightly dampened with a 6 percent bleach solution. After 10 minutes, wipe these items again with a rag dampened with clean water.

   - Seal all used black garbage bags and rags into another bag and put out for normal garbage pickup. Empty the spray can and flush out to remove bleach residue.

10. Remove all clothing worn at the scene, including underwear, and place in garbage bag for laundering and/or dry cleaning (preferably the latter). Take all garbage bags with contaminated clothing to a place where they can be cleaned separately from other garments.

11. All personnel should shower, scrubbing all of the body with soap and water, with particular emphasis on areas around the mouth and nostrils and under fingernails. Shampoo hair and thoroughly clean moustache if you have one.

12. Do not smoke, eat, drink, touch face, or void until step 11 is completed.

13. Put on clean clothes. Place apparatus back in service when decontamination is completed.

14. Have cleaned firehose and SCBA checked by competent personnel before placing it back in service.

15. Arrange for the sterilization bags to be taken to a hospital laundry facility for cleaning and sterilization of the fire coats, gloves, and any other garments sent in.

Reminder

Black garbage bags are to be used for items retained at the station. Orange bags are for items sent away for sterilization.

To Change SCBA Cylinders at the Scene

Flush empty cylinder and surrounding area of fire fighter's back with copious amounts of low pressure water from open end of firehose. Also flush facepiece and breathing tube to prevent inhalation of harmful material when regulator is disconnected.

Wear gauntlet-type rubber gloves, such as those used by linemen, when changing cylinders. Flush gloves after use before removing them.

Place empty cylinder in black plastic garbage bag and seal for subsequent decontamination.

The person doing the flushing and cylinder-changing must wear turnout gear and SCBA.

DECONTAMINATION PROCEDURE
LEVEL R: RADIOACTIVE HAZARDS

At the Scene

1. Preparation

    A. Mark off a decontamination area with two parts.

    B. Make up a solution of detergent and water. Obtain scrub brushes.

    C. Set out a reserve air supply, preferably with a work line unit or otherwise with a spare SCBA.

    D. In the first part of the decontamination area, set up a runoff capturing method, either with wading pools or through the use of tarpaulins.

    E. If appropriate, a "walkway" of polyethylene sheeting (weighed down if necessary) can be laid down from the exit of the incident scene to the decontamination area, to prevent possible contamination of the ground.

2. The decontamination crew will don SCBA and, where available, disposable chemical suits.

3. The fire fighters suspected of being contaminated will be scanned carefully with a radiation monitor suitable for detecting surface contamination. All parts of their clothing and personal equipment will be scanned, including the soles of the boots. If no readings are found, the personnel that have been checked can leave the decontamination area.

4. Personnel found to be contaminated will be scrubbed down thoroughly with the detergent solution by the decontamination crew. This is followed by a flushing off using low pressure water. Efforts should be made to capture the runoff.

5. The fire fighters will then move to the second part of the decontamination area, where they will be scanned again with the radiation monitor. If any readings are found, they will return to the first part of the decontamination area and step 4 will be repeated.

6. When all personnel have been cleaned of contamination, the decontamination crew members themselves will be hosed down. The matter of the captured runoff water will be discussed with environmental authorities and disposal arranged in a manner acceptable to them.

7. In the event fire fighters being decontaminated run out of breathing air, the reserve supply set out in step 1 will be passed to them. They should hold their breath while changing face pieces.

8. In the event that, despite repeated scrubbing, any fire fighters cannot be decontaminated, they will remove as much of their clothing as possible in the second part of the decontamination area, and don clean or spare clothing. The clothing that has been taken off will be sealed into garbage bags and returned to the station. This evolution must be executed in such a manner as not to contaminate the clean clothing.

9. Any equipment suspected or known to be contaminated will be sealed into garbage bags and returned to the station.

On Return to Station

Follow the Level C procedure, steps 5 to 12, for those fire fighters who were found to be contaminated in step 3 above, and for any contaminated equipment.

To Change SCBA Cylinders at the Scene

Personnel emerging from the incident to have their breathing apparatus cylinders changed will be scanned with a radiation contamination monitor in a manner identical to step 3 above.

If no readings are found, the fire fighters can proceed to the SCBA cylinder change area and may then return to the incident with fresh cylinders.

Personnel found to be contaminated may not return to the incident. They will be put through the full Level R decontamination procedure, and other fire fighters will be sent to the incident to replace the fire fighters withdrawn.

Before the replacement fire fighters go in, they should attempt to obtain information as to where the other personnel might have received their contamination, in order to allow them to observe the necessary caution when approaching that area.

Note: Steps 1 and 2 of the Level R procedure must be in place by the time the first fire fighter emerges from the incident. If circumstances permit, these preparations should be made before personnel even enter the incident area for the first time.

Decontamination: Specific Observations

Pre-incident Planning

Review the procedures and, if they are suitable for your location, assemble the equipment necessary into an easily transported container. Some departments, for instance, have all the special items needed for etiologic decontamination carried in a "Level E Decontamination Kit."

Many departments will have infrequent need to use these procedures. To prevent skill decay, and to prevent certain critical steps in the procedures being accidentally left out, it is suggested that a copy of the procedures be available at the scene and that regular training in the procedures take place. Executing these procedures accurately is not as easy as it would seem.

The time when you have twenty garbage bags with contaminated clothing sitting on your apparatus floor is not the time to start looking for a laundry that will clean them. Most commercial cleaning companies will not be interested in handling contaminated clothing.

Furthermore, it should be recognized that at some incidents the nature or extent of the contamination may be such that full decontamination is beyond the resources of the fire department (especially with Levels C, E, and R) and will require specialist treatment. With these three levels, consideration should be given to the destruction of all permeable items in case of serious exposure.

You should therefore make prior arrangements for the following:

- Obtaining steel drums at any time of the day or night. The drums must be clean and must have a removable lid—not just a bung and vent-hole.

- Analysis and expert decontamination of equipment and clothing contaminated by severely hazardous substances. This is needed for Level C and Level R, although different companies are likely to be needed for the two levels.

- Acceptable methods of disposal for items that cannot be cleaned, or that would be uneconomic to attempt to clean, for Levels C, E, and R contaminants.

- The use of a hospital laundry service to perform Level E decontamination on fire gear. This laundry should be approached for the loan of a number of sterilization bags, which are typically used in the hospitals to put dirty laundry in for shipment to the laundry service. Check that the hospital laundry service can take fire coats; in some cases the buckles may bash the inside of their machines too much.

- Check the availability of replacement fire gear and equipment that can be used while the original items are out being decontaminated under Levels C, E, or R.

You may want to establish a policy regarding personal items such as rings, wallets, watches, etc. Many of these, especially leather items, cannot be decontaminated and may have to be destroyed. Fire fighters should be aware of their department's policy with regard to recompense or replacement.

One further item of preplanning will always stand you in good stead—note the names and contact numbers of any local experts who could assist and advise you during the incident and its subsequent decontamination.

Decontamination Area Layout

When choosing the location of the decontamination area, consider the following:

- Prevailing weather conditions (temperature, precipitation, etc.)
- Wind direction
- Slope of the ground
- Surface material and porosity (grass, gravel, asphalt, etc.)
- Availability of water

- Availability of power and lighting

- Proximity of the incident

- Location of drains, sewers, and watercourses

When setting up the area, provide the following features:

- Containment of wash-down water if that is necessary

- Spare supply of breathing apparatus (extra SCBA, extra cylinders, or work line units)

- A supply of industrial-strength garbage bags, double- or triple-bagged if necessary

- Clearly marked boundaries, not just a rope lying on the ground

- Clearly marked entry and exit points with the exit upwind, away from the incident and its contaminated area

- A waiting location at the entry point where contaminated personnel can await their turn without spreading contamination further

- Access to triage and other medical aid upon exit if necessary

- Protection of personnel from adverse weather conditions

- Security and control from the setting up of the area to final cleanup of the site

Environmental Considerations

One fundamental concept forms the basis for these decontamination procedures:

"The human being comes before the environment."

Notwithstanding the above, where containment of run-off is called for, genuine attempts must be made if only to avoid possible legal consequences. Examples of containment basins are listed below:

- Children's wading pools

- Portable tanks (as used in rural fire fighting)

- Tarps laid over a square formed by hard suction hose or small ground ladders

- Diking with earth, sandbags, etc., covered with tarps

Fire fighters stepping out of a containment basin should lift one foot, have it rinsed off so the water falls inside the basin, step out with that foot, and repeat for the other foot.

When the containment basin is full, it should be able to be siphoned or pumped off into drums or into a vacuum truck for controlled disposal in a manner acceptable to the authority having jurisdiction.

Any runoff that is not contained will eventually enter sewers and water courses, or if it sinks into the ground, will ultimately reach the water table. The Department of Mechanical and Fluid Engineering at Leeds (U.K.) University has determined that as long as a chemical is diluted with water at the rate of approximately 2000:1, pollution of water courses will be significantly reduced.

There is also a change in attitude coming with the environmental authorities, whereby they recognize that the small amount of chemical likely to be washed off contaminated fire fighters with adequate dilution will result in minimum damage to the environment, especially when compared to the results of the spill that generally led to the personnel contamination in the first place.

Any substances that enter sewers and water courses should be reported to environmental authorities and to the sewage treatment plant likely to receive it. If necessary, advise water authorities downstream from the decontamination area of actual or potential pollution.

The most appropriate decontamination for materials that have a severe effect on the environment will usually be to use minimal amounts of water, with runoff containment. Other substances should be flushed off personnel with the 2000:1 factor as a minimum guideline.

Weather

If decontamination is done indoors because of bad weather, ensure that the drains go into a holding tank and not directly into the sewers.

If the hazardous material involved requires Level D decontamination, and it is raining or snowing, protect fire fighters from the precipitation until they have been processed.

Take care when using instruments in wet weather. Extreme cold may affect the operating effectiveness of instruments, especially delicate ones originally designed for use in laboratory environments.

Under extreme weather conditions (heat or cold), decontamination personnel must be rotated more frequently.

These decontamination procedures should be reviewed in light of your local climate and adapted if necessary where your area's weather conditions dictate.

Fluid Replacement

At hazardous materials incidents, especially when chemical suits are worn, serious dehydration can occur in fire fighters. Replacement of fluids should only be permitted if at least gross decontamination is performed—a washdown especially around the head and upper body.

The preferable method of consuming liquids is by means of drinking boxes with straws (the straw inserted by someone with uncontaminated hands), or by means of a squeeze bottle with an attached drinking tube as used by athletes.

Chemical Suit Decontamination

When a chemical suit is taken off its wearer, a suitably protected assistant should roll it in on itself in order to keep the outside of the suit from coming into contact with the wearer.

Because of the inherent smoothness and impermeability of chemical suits, it is usually only required that the on-scene washdown part of fire fighter decontamination be performed. Upon return to the station, instead of doing the steps listed in the appropriate procedure, fire fighters should wash and rinse the chemical suits and examine them carefully for damage caused at the incident. Zippers should be lubricated with their special lubricant.

Followup communication with the suit manufacturer as to the exposure, as well as followup from the exposing chemical's manufacturer, is useful in determination of the long-term effect of exposure to chemical-protective ensembles. Any questionable or unusual findings anywhere in the decontamination or testing process should be immediately referred to the manufacturers; the clothing should be placed out of service until it can be repaired or reevaluated.

Vacuum Cleaners for Level D

When selecting a vacuum cleaner, the following points should be taken into consideration:

- Can it operate off a generator, or is it unforgiving so far as voltage fluctuations are concerned?

- Will it operate safely in an area where it might get wet?

- How effective are the filters?

- Can the unit itself be safely cleaned out?

- Are replacement hoses easy to obtain?

Although you won't want to operate your vacuum cleaner under water, it might accidentally get splashed, so some basic water protection will be of benefit.

The degree of filtering achieved is important. Most wet/dry industrial-type vacuums will achieve a reasonable effectiveness. Some specialized cleaners, equipped with HEPA (High Efficiency Particulate Air) filters will go down to 0.3 microns, but they are expensive. The small, cigarette-lighter-plug powered car interior cleaners are not suitable as they filter very little, instead blowing most particulate they pick up back out through their exhaust ports.

Easy removal of contaminated filters will help, as will good access to the machine's insides for its own decontamination. You will usually find that it is impossible to guarantee the effectiveness of cleaning the accordion-style hose, and you should consider replacing these if they become contaminated.

Remember not to operate a vacuum cleaner in a flammable or explosive atmosphere, unless yours is intrinsically safe.

Bleach

Bleach, as shown on the bottle's label, is corrosive. Do not spray bleach on fire fighters' skin—it hurts! It is also reactive—do not let it come in contact with fuels and solvents, as a heat-generating reaction (and possibly, fire) will result.

Do not spray bleach on bunker gear as it deteriorates and discolors the garment. It also impairs its fire retardancy.

Regular bleach containers are often made of hard plastic and are liable to crack or leak around the cap when carried on a vehicle. Consider transferring the bleach to a container such as a new, unused plastic gasoline can, which is far sturdier, but then be sure to relabel the can appropriately. Never put bleach in a metal container as it will react with the metal.

At room temperature, bleach shielded from sunlight will degrade about 1 percent per year, i.e., from 6 percent to 5 percent (faster in warmer temperatures, slower at colder temperatures). You should therefore replace the bleach at the appropriate time with a fresh supply, as its strength will have decreased with time.

Recordkeeping

A member of the crew responsible for performing the decontamination should maintain written records of the following:

- Fire fighter's name, material involved, length of exposure
- Level of decontamination performed
- Any ill effects observed
- Where the fire fighter went after decontamination, i.e.:
  - ✓ Returned to station
  - ✓ Sent to rest area
  - ✓ Removed to hospital
  - ✓ Reassigned to other duties at the scene, etc.

At the station, entries should be made on the fire fighters' medical records of the incident date, material involved, and decontamination performed, where exposure is known or suspected. This will assist both in tracing future sickness through synergistic effects of chemicals in the body and with support of any later injury or sickness claims.

If appropriate, records should also be kept of the length of time each chemical suit was exposed, and what substance it was exposed to. This will permit the tracking of cumulative degradation of the suit material due to exposure to a variety of chemicals or due to repeated exposure to one particular substance.

Contamination of Vehicles

Any vehicle driven through a contaminated area must be washed down, including the undercarriage, chassis, and cab. Air filters on vehicle (and, where appropriate, generator) engines must be replaced. Porous items such as wooden hose beds, wooden equipment handles, seats, and cotton jacketed hose may be difficult to clean completely and may have to be discarded.

It is therefore better to take the "uphill and upwind" approach and keep vehicles at a suitable distance from incidents.

One Final Observation

The entire foregoing contents of this document have probably made you realize by now that it is much more desirable to handle hazardous materials incidents with chemical suits than with regular fire fighting turnouts. The cost of disposable suits is relatively cheap; even for a small department, throwing away a few hundred dollars worth of disposable suits after one single use will be cheaper than replacing turnouts or paying for commercial cleaning. In many jurisdictions, the fire service is permitted to recover the cost of destroyed equipment from the party responsible for the incident's occurrence, and the cost of disposable chemical suits can thus be recovered.

Always remember: if the emergency response crew is not equipped with gear suitable for entry into a hazardous or toxic atmosphere, then the option of "no go" should be considered the most appropriate tactic.

Speed: A Case for Exception?

Decontamination should emphasize thoroughness, not speed. Under noncritical conditions certain common sense actions should be taken, such as decontaminating the fire fighter with the lowest air reserve first.

Speed is only important where a victim is involved and even then decontamination should be as thorough as is practicable.

Circumstances may dictate that emergency decontamination becomes necessary, such as when a protective suit has become split or damaged, or when a fire fighter is injured. Emergency decontamination may also be applicable when contaminated civilians or other emergency workers (police, ambulance, etc.) are involved.

Emergency Decontamination Procedure

Paragraphs 1 to 6 below, although arranged in a basic chronological order, do not necessarily have to be undertaken in the exact sequence outlined. The officer-in-charge should act in the most expedient manner appropriate without worsening the situation.

The procedure outlined should be carried out as quickly as possible.

To protect the ambulance crew and hospital staff as well as the victim, every attempt must be made to perform at least this emergency procedure prior to transporting the victim to the hospital.

1. Remove the victim from the contaminated area into the decontamination zone and ensure he or she is supplied with uncontaminated air or oxygen.

2. Remove fire helmet, if worn, and immediately wash with flooding quantities of water any exposed parts of the body that may have been contaminated.

3. If the victim is wearing SCBA, release the harness and remove the set leaving the face mask in position.

4. Remove contaminated fire gear or clothing (if necessary by cutting it off the victim) ensuring, where practicable, that the victim does not come into further contact with any contaminant. Maintain the washing of the victim while the clothing removal is taking place.

5. Remove the victim to a clean area. Render first air as required, but do not apply mouth-to-mouth resuscitation. Send victim for medical treatments as soon as this emergency decontamination procedure has been completed.

6. Ensure hospital/ambulance personnel are informed of the contaminant involved.

Index